D0904846

WHO OWNS THE DEAD?

Who Owns the Dead?

THE SCIENCE AND POLITICS OF DEATH AT GROUND ZERO

JAY D. ARONSON

HARVARD UNIVERSITY PRESS
Cambridge, Massachusetts, and London, England
2016

Printed in the United States of America
First printing

Library of Congress Cataloging-in-Publication Data
Names: Aronson, Jay D., 1974– author.
Title: Who owns the dead? : the science and politics of death at Ground Zero / Jay D. Aronson.
Description: Cambridge, Massachusetts : Harvard University Press, 2016. |
Includes bibliographical references and index.
Identifiers: LCCN 2016011691 | ISBN 9780674971493 (alk. paper)
Subjects: LCSH: Mass casualties—Identification. | Forensic anthropology. | Terrorism victims' families—United States—Attitudes. | Memorialization—Political aspects—United States. |
September 11 Terrorist Attacks, 2001—Political aspects.
Classification: LCC RA1055 .A76 2015 | DDC 614/.17—dc23
LC record available at http://lccn.loc.gov/2016011691

Contents

WHO OWNS THE DEAD?

Introduction

The 2,753 victims of the September 11, 2001 attacks on the World Trade Center in New York City were business people, lawyers, janitors, bond traders, electricians, secretaries, food service workers, firefighters, police officers, engineers, computer specialists, mothers, fathers, sons, daughters, brothers, sisters, cousins, lovers, friends, spouses, and community members. The remains of 1,113 of them have not been identified. Of those who were found, all but 293 were recovered from among 21,900 bits and pieces scattered throughout the debris of the fallen towers: a tangle of steel beams, rebar, pulverized concrete, asbestos fiber, plus the contents of thousands of offices and retail outlets.[1] Buildings that were once 110 stories collapsed into the space of just eleven, seven of which were below street level. Rescue workers initially picked through the rubble by hand, frantically searching, first for survivors and then for victims' remains. They were choked by dust and smoke and the stench of death. Soon, giant bulldozers and grapplers took over the job of removing debris, and rescue workers dedicated themselves solely to finding remains, including their own brethren.

The World Trade Center was attacked just as large-scale DNA identification efforts were becoming possible. The biotechnology boom of the 1990s had produced technologies that could be used to rapidly extract and analyze genetic material from biological specimens. Simultaneously, scientists involved in the investigations of large-scale accidents and mass atrocities were learning how to apply these tools to the damaged and degraded forensic specimens recovered from complex graves. Human rights advocates and activists also realized that families benefit from the return of their loved ones not just spiritually and psychologically, but also socially and legally. Without identification of their loved ones, relatives cannot access financial and social services, dispose of personal property, or seek compensation for their loss. They can also become socially marginalized.[2]

These advances led New York City's chief medical examiner, Charles Hirsch, to promise that he and his staff would attempt to identify and return to families every human body part recovered from the site—even those that were heavily damaged by the collapse of the towers and the underground fires that raged at the site for weeks. The job would not be easy—it would require a bewildering mix of technological expertise, statistical acumen, and persistence.

More than $80 million has been spent on the effort thus far, and the Office of the Chief Medical Examiner has committed to continuing in perpetuity the effort to identify remains as new techniques become available. The primary goal, of course, is to link even the tiniest fragment of human remains to a person in an effort to provide proof of death for those families that hunger for such knowledge.[3] But the massive forensic effort was also undertaken to demonstrate that Americans, as individuals and as a society, were dramatically different from the terrorists who so callously disregarded the value of life. It was as much a political and moral statement as it was a scientific and legal one.[4]

This is not a book for the faint of heart. It tells the story of the recovery, identification, and handling of human remains in the aftermath of the September 11, 2001 terrorist attacks on the World Trade Center. It also delves into the contested efforts to memorialize the victims of the attacks both at the World Trade Center site and at the Fresh Kills Landfill, where much of the debris from the Trade Center was taken for sifting and disposal, and the controversy over the storage of remains at the National 9/11 Memorial and Museum. It exposes the raw grief and persistent anger that motivated a small group of families to continue to contest redevelopment and memorialization efforts at the site more than a decade after the 2001 attacks. In addition, this book seeks to explore the impact and legacy of efforts to recover, identify, and memorialize the dead on the families of victims, the City of New York, the nation, and the world.

September 11 was the first time since the Civil War that such a large number of dead bodies had to be dealt with on American soil.[5] Yet the United States had a history with the issue outside its borders: its complicity in dissident disappearances in Latin and South America during the 1970s;

in lending scientific expertise to the identification efforts in those same countries in the 1980s and early 1990s; in leading the international effort to identify the missing after the Balkan wars of the 1990s; through its efforts to recover the remains of American soldiers missing in foreign wars; and in the blatantly political efforts to uncover mass graves in Iraq in order to justify the invasion of the country in 2003.

Analyzing the U.S. response to mass death can help Americans better understand similar events around the world, and to empathize with people confronted with such atrocities. Global policy cannot be developed based on the uniquely American response to 9/11, but the United States can no longer turn a blind eye to the psychosocial, political, and scientific needs of societies struggling to cope with mass death. Policy makers can no longer assume that locating bodies and reburying them is enough—the World Trade Center story amply demonstrates that the exhumation and identification of human remains is inherently political and fraught with controversy from beginning to end.

Human remains have political, cultural, and emotional power.[6] The death of a loved one in a mass disaster or an act of terror can leave relatives of victims and the missing feeling emotionally and spiritually drained. But it can also give them special status within society and a voice that can be used to make demands on government institutions that would ordinarily not listen to them.[7] Emboldened relatives of the dead and the missing, especially women, often speak out on social and political issues with little regard for negative repercussions. They are fighting for justice and the return of their flesh and blood, and little else matters to them. This is true of mothers, wives, and grandmothers of victims of war, disaster, and mass killing around the world.[8]

Since 1977, for instance, the Mothers of Plaza de Mayo don white scarves and march in the center of Buenos Aires every week to demand information about, and accountability for, their children who went missing during the 1976–1983 military junta in Argentina.[9] Similarly, in the aftermath of the 1995 genocide in Srebrenica, family groups in Bosnia successfully demanded that international actors identify and return bodies of their missing loved ones rather than just gather demographic profiles for use in war crime prosecutions.[10] New York City witnessed the same situation after September 11, when wives, sisters, and mothers became advocates for the missing and the dead. Many men also became advocates for the victims of 9/11—especially firefighter fathers searching in the rubble for their firefighter sons.

In addition to the actions of relatives left behind, mass death necessitates broader social, political, and cultural responses. Families and communities look to honor the dead and, in some cultures, ensure their smooth transition from the realm of the living to the realm of the dead.[11] The state hopes to reassert its control over society, especially when it played a role in—or failed to prevent—the disaster. It seeks also to demonstrate care and concern for the lives of its citizens and the nation as a whole.[12] For everyone involved, and especially forensic scientists, there is a more general desire to ensure that the violation of the dead does not remain permanent.[13] While forensic identification cannot bring the dead back to life, it can restore some sense of normalcy to families and communities whose loved ones have died in traumatic and violent ways.

Individual Identification and Collective Commemoration

In many places, including the United States, violent mass death can stigmatize the location where it occurs, and the site must be cleansed, destroyed, or transformed into a memorial.[14] There is also a practical problem: what to do with the bodies and body parts, particularly when a substantial portion cannot be identified and returned to families? In the case of the September 11, ownership and control of unidentified remains greatly affected debates about the future of the sixteen-acre World Trade Center site.

Thus, beyond the forensic dimensions of identifying the missing, this book explores how human remains became central to the memorialization process at the World Trade Center site. The mere possibility of eventual identification means that the bones can never be buried and forgotten. Instead, they must be maintained in an active repository, keeping both the remains and families in a state of extended limbo. We are only beginning to address this dynamic, the salience of which has increased dramatically in the age of DNA identification.

After the attack on the U.S. naval base at Pearl Harbor on December 7, 1941, the vast majority of the 1,177 navy and marine personnel who were killed on the battleship USS *Arizona* were classified as buried at sea and left in place underwater. While the victims could have been recovered—the ship rested close to shore and its parts and materials were salvaged throughout the war—they were assumed to be unidentifiable due to their fragmented and burned condition.[15] In the 1995 Oklahoma City bombings, human remains that went unidentified were referred to as "common tissue" and collectively buried in a memorial tree grove near the state capital.[16]

After the 2001 World Trade Center attacks, on the other hand, each remain was stored separately and treated as an individual entity that might one day be identified as new forensic techniques became available. This policy ruled out collective interment. It also meant that the creation of a "tomb of the unknowns," such as had become popular after World War I in Europe and the United States, would be unlikely.[17]

Historian Thomas Lacquer argues that several factors led to a shift in Western (and particularly Anglo-American) conceptions of what ought to be done for the victims of violent death, especially in combat. Prior to the twentieth century, war victims were generally left in place to be eaten by scavengers or buried in mass graves.[18] By World War I, there was a concerted effort to bury them individually in marked graves and to memorialize them by inscribing their names on grand monuments after the war. One reason for this change was that, by the twentieth century, soldiers were fighting on behalf of democratic nations and were thought to deserve equality of treatment in death as in life.[19]

But democracy and politics on their own are not sufficient explanations for this change. Lacquer argues that the sheer magnitude of the slaughter in the trenches demanded a new response. The Great War monuments attempt to make sense of the scale of deaths, while making manifest their ultimate incomprehensibility. Perhaps most poignantly, Lacquer notes that many who fought in the battles, as well as commentators who wrote and spoke about the Great War, were worried that the sacrifices of these young men would soon be forgotten and all evidence of their deaths would be subsumed by nature retaking the land. "There is evident here a powerful anxiety of erasure, a distinctly modern sensibility of the absolute pastness of the past, of its inexorable loss, accompanied by the most intense desire to somehow recover it, to keep it present, or at least to master it."[20] The fears were compounded by the condition of remains on the battlefield that resulted from the new machinery of war. Shelling, land mines, artillery bombardments, and machine guns produced not complete bodies with bullet or stab wounds, but mounds of flesh, and disarticulated arms, legs, torsos, scalp, blood, and tattered uniforms. In other words, if the dead were not buried as individuals and their names were not recorded on massive monuments, then they might disappear into the soil and be forgotten.

Nearly a hundred years later, this fear of erasure seemed to motivate at least some of the families of the World Trade Center victims and their allies. In the aftermath of World War I, nature would reclaim the battlefield and render the events that took place there—and those who died there—invisible. This time, redevelopment would be the culprit, as well as the desire of city residents and city leaders to put the horrible events of 9/11 behind them and get on with business and life. In many ways, this tension between the desire to memorialize and remember, and the desire to move on would animate debates about the site for much of the next decade.

Further, new genetic technologies have changed the way we remember the dead. The emergence of DNA identification means that it is far less likely that there will be unknown soldiers in future wars. As a result, monuments to unknown and unnamed war dead will no longer be a way to honor their sacrifices.[21] Similarly, in the era of DNA identification—and in keeping with the nineteenth century belief that dying an anonymous death and being buried in an unmarked grave was a sign of social exclusion and despair—it is no longer enough to produce a single collective memorial to ordinary people killed in mass conflict events.[22] We are compelled to remember them as individuals and push technology to its limits to identify their remains.

Yet, just as collective memorials and tombs of the unknown served to tie the nation together in the past, these individual stories—underwritten by DNA identification—now serve as conduits for collective understanding of conflict as they become threads of a collective tapestry.[23] They have social and political power that can be called upon when needed. An important motive for memorializing the dead at the World Trade Center is to justify the military and diplomatic actions taken to protect the United States and its citizens from similar attacks in the future. To downplay the human toll of terrorism is to lessen Americans' willingness to put up with war and intrusions on their civil liberties. In a nation that exalts individualism, knowing the names, faces, and stories of each of the victims makes it hard for citizens to accept inaction. While humans are generally able to shrug off the deaths of thousands of strangers—we do it every day while reading, watching, or listening to the news—it is difficult to ignore the death of even one person we have come to know as an individual.

The challenge of memorializing the victims of 9/11 as individuals is that narratives stressing communal sacrifice cannot be made too overtly—visitors, viewers, or readers of these memorial efforts must arrive at this conclusion without noticeable coercion. When the link between individual and nation is made too obvious in the context of 9/11, many victims' families and other stakeholders have protested and actively intervened—especially when the events of 9/11 were used by the Bush administration to justify the war in Iraq and policies that many believed further eroded Americans' privacy and civil liberties, and by activists to shine a light on what they saw as the global struggle for freedom and liberal values.

Ironically, though, efforts by relatives and stakeholders to depoliticize the victims of 9/11, and the memorialization process as a whole, have served to reinforce the use of these victims for political purposes. For example, when the *New York Times* published "impressionistic sketches" of the victims of the World Trade Center attacks in its "Portraits of Grief" feature, it sought to portray them all as living the prototypical American dream or on the way to achieving it.[24] The *Times* editors responsible for these portraits were fully aware of what they were doing. "We recognize," they wrote in an unusually candid commentary on the "Portraits" project on October 14, 2001, "the archetypes that define the ways these stories are told. The tales of courtship and aspiration, the ways these people relaxed and how they related to their children—these are really our own stories, translated into a slightly different, next-door key."[25] Such representations not only erased the diversity of the victims, but also elided the fact that many of them were not U.S. citizens, including at least a few who were undocumented immigrants. The ways in which the *New York Times*—and so many other voices in American culture—used the events of September 11 to tell "our own stories" about America, good and evil, and right and wrong, made it impossible to see the lives of the victims through anything other than a political lens.

For different reasons, investigators also sought to identify the remains of the suspected perpetrators of the attacks. The presence of remains at the crash site would solidify their connection to the crime. Further, families were adamant that the remains of their loved ones not be comingled with those of their murderers—and demanded that these remains be separated as much as possible.[26] For the U.S. government, there was also the more vexing question of how to deal with the identified remains of the hijackers, who used their bodies as weapons and actively rejected the set of international norms and laws that govern conflict among states, including the disposition and treatment of enemy remains.

Such decisions simultaneously invoked law, conceptions of punishment, obligations to victims and their families, the projection of the country's image to other nations and other would-be terrorists, and emotion.[27] The handling of the perpetrators' remains had to be done in a way that neither glorified them nor treated them with the level of respect accorded their victims. Yet there are still no defined policies for dealing with this challenge. After the killing of Osama Bin Laden by Navy Seals, for instance, the U.S. government decided to dispose of his body in the ocean in accordance with Islamic practices when burial on land is not possible.[28] American officials noted that they could have handled the body in a less respectful way—for instance, by dumping it from a helicopter without washing the body or wrapping it in a white sheet—but they determined that a proper burial was the right thing to do. While Muslim scholars disputed the propriety of the effort—burial at sea is generally reserved for individuals who die at sea and cannot be brought to land—at least one called the decision "pragmatic."[29]

In the context of the September 11 attacks, this responsibility involves long-term, or perhaps permanent, storage, because, while technically permitted to do so, neither the perpetrators' countries of origin nor their families were willing to claim their remains once they had been identified. For foreign governments (Saudi Arabia, Egypt, and the United Arab Emirates—all U.S. allies), claiming the remains would be tantamount to admitting that one of their citizens was responsible for the murder of 2,753 people. And in many ways, the decision to become a violent jihadist is a de facto rejection of belonging to any one country in favor of joining the community of believers who answer only to Allah.[30] For families, claiming the remains would be an admission that their kin was indeed a terrorist. As such, the remains of the perpetrators are interred separately from victims' remains in an undisclosed location under the control of the Office of the Chief Medical Examiner (OCME).

Controversy

While this book focuses extensively on the controversies that emerged over efforts to recover, identify, and memorialize the victims of the World Trade

Center attacks, I do not wish to suggest that controversy itself is bad—in fact, controversy is perfectly normal in this situation.[31] The intersection of personal pain, anger, and grief with commerce, real estate, and politics ought to provoke debate and disagreement in a democratic society. To think otherwise would be, as historian Edward Linenthal writes, a "strange assumption."[32]

Controversies give us a window into what people care most about and how things might have worked out differently. They also help us understand how cultural meaning and memory are produced around painful events of the past.[33] By studying controversies as they occur, we can see how disagreements were—or were not—resolved, and how and why certain groups continue to contest the matter at hand after the dispute has been formally resolved. We can also see why these groups are occasionally successful in reopening debate, and why they usually are not.[34]

Ownership

The desire of local and national leaders to rapidly repair the hole in lower Manhattan's fabric was challenged by the multiple meanings of the site. For rescuers, it was a disaster area to be tamed; for military and national leadership, it was a the site of an enemy attack; for residents, it was a shocking and traumatic violation of their homes, communities, and everyday lives, not to mention an environmental nightmare; for families, it was a place of mourning, where a loved one breathed his or her last breath, and a cemetery; for the public, it was a place of absence where buildings once stood and lives were once lived, a place of protest, a tourist site, a place of national trauma, a (re)construction site, a neighborhood, a commercial district, and the center of global capitalism; for the architecture community, it was an opportunity to make an architectural and urban planning statement while remembering the victims and revitalizing lower Manhattan.

The presence of so many stakeholders with so many agendas meant that the World Trade Center became a new kind of battleground: an economic, legal, and moral one over who could claim ownership of the site.[35] At the most basic level, there were significant contractual and financial battles over proprietorship and control of the property. More conceptually, there was strong disagreement about the overall uses to which the site should be put (in which authority was based on expertise, whether professional or lay, and asserted by urban planners, architects, residents, or business owners). Finally, there were ethical debates about what ought to be done to honor and respect the thousands of lives lost on September 11 (in which authority was moral, political, and nationalistic and asserted by all parties, but most forcefully by the families of victims and their advocates).

Sacred Space

Soon after the dust and smoke settled, the question of whether the World Trade Center site would be preserved as a sacred space or brought back to life as the heart of a vibrant, revitalized neighborhood came to the fore.[36] For those who did not see the site as inherently sacred, the best option was to clean up, rebuild, and get on with living. For those who did, the site had to pay homage to the victims of the attacks and could not simply be redeveloped as if nothing had happened there. Ultimately, the design and planning of the redevelopment of the site was about balancing the two—and most of the disputes about the remains revolved around the extent to which the activist families believed the planners did or did not recognize the sacredness of the site.

So, in what sense can the sacredness of the World Trade Center be understood? Religious studies scholar David Chidester and historian Edward Linenthal highlight two broad theoretical frameworks that can be used to answer this question. One argument holds that sacredness is physical and emerges from the place itself—either as a result of events that happened there, or from some essential property of the site that projects power or spiritual qualities that elevate it above other locations. This is the view held by many family members who lost loved ones on 9/11. The other argument, which emerges especially from the work of French sociologist Émile Durkheim, is situational.[37] In this view, sacredness is not inherent but is produced through human action for specific social ends. The sacred is produced through the cultural work of sacralization.

Chidester and Linenthal go on to argue that there are three main characteristics of sacred space: it serves as a place for rituals, which they define as formalized, repeatable symbolic performances; it causes visitors to focus on questions of what it means to be a particular kind of person in a meaningful world; and, finally, its power makes it socially valuable and the subject of contestation. Control of sacred space thus becomes an exercise of power. Indeed, a key aspect of the story told in this book is that of families of victims who were fighting to preserve a space for their loved ones' memories, free from other interpretations or stories of anyone else's suffering. At the same time, other stakeholders sought to erase what happened, or at least to provide alternative meanings that enabled life, and commerce, to resume at the site. In the end, a sort of compromise was reached and the World Trade Center site was divided into gradations of sacredness: the OCME repository was completely sacred; the memorial and museum less so but still partly sacred; and the rest of the site was profane, but still requiring some degree of reverence and respect.

Museum studies scholar Paul Williams invokes the concept of "secular sacredness" to discuss memorialization, precisely because of the complex, ambivalent nature of these remains and the numerous religious understandings of the "sacred" that exist around the world. In contexts like the World Trade Center site, traditional notions of religious sacredness break down for many reasons, not least of which is the fact that visitors come from numerous religious traditions, and that one religion or another is also implicated in the atrocity being remembered. Even when this is not the case, religion rarely provides an acceptable explanation for why people commit inhuman acts. Further, most memorial museums ultimately put forth a universalist vision that all human lives have value and that secular, rational support for human rights is our best hope for a more peaceful and just future.[38]

Williams notes the often-contradictory effects of displaying and/or housing remains within memorial museums. While the presence of human remains signals the importance of the site and enables rituals regarding the dead to take place (such as pilgrimage, prayer, and mourning), the display of these remains may render them profane by preventing traditional (and religiously bound) funerary rites. There is a sense in which the remains lend the site authenticity and connect visitors to the tragedy that happened there, but that the use of remains in this way can further reinforce their profane condition.

Ultimately, the 9/11 Memorial Museum decided to thread this needle by storing the remains out of view of the public with the public's knowledge. The designation of the repository as a "separate space" walled off from the memorial museum denotes both its power and importance to the site overall (in that the remains make it impossible to deny that death occurred there) and the potential the remains have to taint the site in some way if not handled appropriately. This placement became a contentious issue for many victims' families. They felt that their loved ones, or the loved ones of other families in their position, would become a museum exhibit and therefore be debased, or rendered mere objects.

Nationalism

Complicating matters even more is the nationalist dimension of 9/11. While the victims of 9/11 may not have died in direct service to the nation, and many of them were citizens of other countries, their deaths took on a broader meaning for all Americans because they were targeted by enemies of the United States. When the city opened a viewing platform at Ground Zero at the end of December 2001 to allow the public to see the progress being made in the Pit and to pay tribute to the victims of the attacks, and to bring tourists back to lower Manhattan, outgoing mayor Rudy Giuliani promised visitors a moving experience: "This gives you all kinds of feelings of sorrow and then tremendous feelings of patriotism. I really urge Americans to come here, and everybody to come here, and say a little prayer and just reflect on the whole history of America and how important democracy is to us."[39] Further linking the site to notions of patriotism and sacrifice, Giuliani noted that that denying the public the right to view Ground Zero would be like "denying people access to other sites of [national] historic significance, like Gettysburg or Normandy."[40]

It was in this vein that New York governor George Pataki decided to recite the Gettysburg Address, the speech that Abraham Lincoln delivered in 1863 dedicating a cemetery for Union soldiers killed in the Battle of Gettysburg. In introducing Pataki, New York City mayor Michael Bloomberg noted that "139 years ago President Abraham Lincoln looked out at his wounded nation as he stood on a once beautiful field that had become its saddest and largest burial ground. Then it was Gettysburg. Today it is the World Trade Center, where we gather on native soil to share our common grief." In these two sentences, Bloomberg situated the victims of the World Trade Center attacks within what Linenthal describes as the "comforting narrative of patriotic sacrifice"—the notion that freedom has a price, and the individuals who died on that day sacrificed their lives for all of us.[41] Pataki then recited the speech verbatim: "Fourscore and seven years ago our fathers brought forth on this continent a new nation, conceived in liberty and dedicated to the proposition that all men are created equal. Now we are engaged in a great civil war, testing whether that nation or any nation so conceived and so dedicated can long endure."

Invoking the definition of sacred space as a place that is inherently meaningful because of what happened there, Pataki continued, "We are met on a great battlefield of that war. We have come to dedicate a portion of that field as a final resting-place for those who here gave their lives that that nation might live. . . . But in a larger sense, we cannot dedicate, we cannot consecrate, we cannot hallow this ground. The brave men, living and dead who struggled here have consecrated it far above our poor power to add or detract. The world will little note nor long remember what we say here, but it can never forget what they did here." Pataki finally read the speech's conclusion, which demands that the soldiers who perished on behalf of the Union not die in vain and that "this nation under God shall have a new birth of freedom, and that government of the people, by the people, for the people shall not perish from the earth."

The decision to recite the Gettysburg Address was seemingly an effort to rise above politics and to place the dead at the forefront of the commemoration, yet its inclusion in the ceremony suggests that state authorities felt obliged to acknowledge the dead in some public way without overtly making political capital out of the event. It is interesting that the common grief and national sadness was forcefully articulated not by President Bush or other national leaders, but by local politicians with grand aspirations—Bloomberg, Pataki, and former mayor Rudy Giuliani. In this way, the nation was brought together in a shared sense of grief to form a political community of mourners.[42] This community was not always tolerant, however. Individuals and institutions with actual or suspected ties to Islam or the Middle East became targets of harassment and violence.[43]

Thus, the dead were not remembered in a politically neutral manner. Linking the World Trade Center attacks to the bloodiest battle that had ever taken place on American soil had serious political ramifications. It placed the World Trade Center attacks among the most important events in the history of the United States. It also signaled that the United States, as a nation, was engaged in a battle between good and evil in which the future of freedom and the democratic way of life were at stake. And, given that the core of the Gettysburg address is an explicit acknowledgment of the sacredness of the Gettysburg site, Pataki was formally stating that Ground Zero was in a very real sense hallowed ground—a place where patriots sacrificed their lives for the nation.

Yet this decision was odd in many ways. The Civil War pitted Americans against one another, with the very survival of the nation at stake. The

9/11 perpetrators were an amorphous group of terrorists who belonged to no one particular country—thus the United States was not at war with itself, another nation, or with any one clearly defined entity. What's more, the victims at Gettysburg were soldiers who went into battle knowing that they stood a good chance of death or injury. Except for the uniformed service personnel who rushed into the twin towers, the victims of the World Trade Center attacks were civilians in every sense of the word. They had gone to work that sunny, warm morning expecting to do their jobs and then return home at the end of the day to family and friends. They had no intention or expectation of dying, and certainly were not representing the nation in any way other than as ordinary citizens. Yet, none of that seemed to matter to those hungry for retribution. The rhetoric and actions of politicians made them into de facto martyrs whose lives, and deaths, needed not just to be remembered and honored, but also avenged.[44]

Further, Lincoln went to Gettysburg to consecrate a cemetery for the dead. One year after the World Trade Center attacks it was clear that the site would be neither a cemetery nor principally a memorial to the victims of the attacks. Unlike the battlefield at Gettysburg, the World Trade Center was not a placid piece of farmland. It was an urban center, and not just any urban center. The World Trade Center was at the heart of the nation's largest and most important city—capital of the arts and finance, a canyon of majestic skyscrapers, the home of the Statue of Liberty and the symbolic font of America's rich immigrant tradition, and generally the place that defines what it means to be successful in the United States. It was also home to eight million people who were fiercely proud of being New Yorkers. There were simply too many stakeholders involved for the stated sacredness of the site to last. At Ground Zero, we can see the interests and dignity of the dead (and their families) clashing with broader local, national, and global interests.

This tension is evident in the progressive reduction of space that was devoted to commemorating the victims of the attacks. Mayor Giuliani and many family groups initially argued that the entire sixteen-acre site ought to be devoted to remembering the victims, but most of them quickly settled for eight acres. Soon, "sacred ground" became limited to the "footprints" of the towers. According to cultural studies scholar Marita Sturken, "The idea of a building's footprint evokes a sense that a structure is anchored in the ground. It is also anthropomorphic, as it implies that the building left a trace, like a human footprint, on the ground."[45] This notion that the space where a building once stood is a suitable place to mourn the dead is also the seen in the Oklahoma City memorial. "The emphasis on the footprints of the two towers demonstrates a desire to situate the towers' absence within a recognizable tradition of memorial sites. The idea that a destroyed structure leaves a footprint evokes the site-specific concept of ruins in modernity. In the case of Ground Zero, one could surmise that a desire to reimagine the towers as having left a footprint is a desire to imagine that the towers left an imprint on the ground."[46]

Other than the presence of Mayor Bloomberg, former mayor Rudi Giuliani (who began the name reading), the governors of New York and New Jersey, and the regional accents of many of the people who read names, allusions to New York City were notably absent. The rest of the more than two-and-a-half-hour anniversary ceremony consisted of dignitaries, survivors of the attacks, and relatives of the victims of the attacks reading the names of the 2,753 people who perished in the attacks on the World Trade Center. In the background, a string ensemble, including cellist Yo-Yo Ma, played wistful, mournful music. The pain in the voices of the readers was palpable and one can see the grief in the faces and bodies of families who were in the Pit. The scene remains nothing less than heartbreaking. Only one other speech was given that day: New Jersey governor James E. McGreevey closed out the ceremony by reading excerpts of the Declaration of Independence. Thus, at least at the one-year anniversary of the attacks, 9/11 was framed in terms of martyrs and a threatened democratic nation, reinforcing the notion that the deaths of the victims (even foreigners and the undocumented) was a patriotic sacrifice and not a random act of murder—and that the entire nation belonged to a single community of mourners united by common grief (for the victims), common principles (democracy and freedom), and common purpose (to avenge the deaths of the victims, to defend our principles, and to defeat the terrorists and terrorism itself).

Grief and Mourning

The aftermath of 9/11 also made plain the reality that grief is both an individual and a collective phenomenon which, contrary to popular belief, does not operate on a regular schedule. While some people pushed their lives forward in the aftermath of 9/11, others remain focused on the events more than a decade after their occurrence. One of the goals of this book is to understand the actions of the small group of families that have remained active in seeking to influence the memorialization of the victims and the redevelopment of the World Trade Center site more than a decade after the attacks. It is tempting to pathologize the actions of these individuals, first dubbed the "memorial warriors" and later the "grief police" by *New York Magazine*, to make it seem as if they have not "moved on" from their loss, or that they are seeking to gain control of the rebuilding and memorialization process as a way to compensate for their inability to bring back their loved ones.[47] To do so, however, would be to assume that that their demands are unreasonable and that those in power have truly made a good-faith effort to accommodate their needs and desires.

Numbering no more than a few dozen a decade after 9/11, this group was made up primarily of middle-class wives, mothers, and sisters of victims and firefighter fathers of firefighter sons, almost all from the New York metropolitan area. This group self-consciously refused to relinquish their ownership claims or moral authority over the human remains associated with the World Trade Center attack victims and indeed the site itself. They also claimed to speak for a sizable population of 9/11 families that agreed with them but were unable to speak out publically on 9/11 matters for reasons of emotion, family duties, economic hardship, or geographic distance from New York. They also seem to have been ordinary people before 9/11 who had previously shown little desire to be public advocates. Loss, anger, and a feeling that the Bloomberg administration and the Lower Manhattan Development Corporation (LMDC) simply didn't care about them transformed them in fundamental ways. "I was a different person before 9/11," Sally Regenhard, mother of firefighter Christian Regenhard, told journalist Deborah Sontag. "I tried to speak out, let's say in Co-op City, where I lived. But now—I'm fueled by adrenaline, outrage and love for my son and that has made me a bigger pain in the ass than I ever was before."[48]

How should we try to understand their activities and the intense mainstream media interest they received? Linenthal highlights four narratives that came to predominate after the Oklahoma City bombings: a progressive narrative (in which the city recovered from the event and came back stronger than ever), a redemptive narrative (that God had a plan for those who died and those who lived), a toxic narrative (that the city had suffered a great trauma from which it would never fully recover), and a traumatic narrative (that the city had suffered a great trauma from which it had to recover).[49] These four narrative structures have many parallels in the context of 9/11, with the addition of a fifth narrative structure that focused on the political dimensions of the attacks and the emergence of a long, global war against terrorism.

For Linenthal, the Oklahoma City bombing is an "unfinished bombing," in that it continues to "claim people through suicide, to shatter families through divorce, substance abuse, and the corrosive effects of profound and seemingly endless grief. It is a toxic narrative, and it exists alongside of, and intermingled with, the other story lines."[50] Despite the existence of this toxic narrative, Linenthal notes that "there seemed throughout the city—indeed throughout the culture—an unspoken statute of limitations on mourning. The failure to 'get on with it' or 'be back to your old self' after a prescribed period indicated, according to the traumatic vision, the presence of an illness and the need for treatment."[51] In other words, grief and mourning were defined by what was considered socially acceptable. This included when, how, and for what time it was appropriate to be despondent, which types of behaviors were normal and which were not, and when one should be done with the mourning phase and transition to the getting on with life phase. Linenthal notes that this perspective is most consistent with the regenerative and redemptive narratives, but only the toxic narrative encapsulates the reality of chronic affliction and an inability to 'put the past behind you' that so many of the bereaved felt. The toxic narrative suggests not a return to the old self or putting one's life back together, but rather a shaping of a new self in the aftermath of such an experience.

Linenthal demonstrates that there were strong differences of opinion within the bereaved community about how public grief ought to be. For some families, it was a private matter that was nobody's business. Others, however, wanted the world to know who their loved ones were and offered a public eulogy by speaking to the press, making the funeral open to the media, or both. Those who opened up to the media gave voice to a kind of communal grief. Other families responded by retreating into the woodwork, while still others took on outspoken, often strongly political, activist roles, advocating for particular forms of memorialization and remembrance or changes in public policies regarding the legal system, victims' rights, and the death penalty.

Whatever the case, it is clear that all of the people who were killed in Oklahoma City and on September 11, 2001 died "culturally significant public deaths," setting them apart from the thousands of people who die every year from everyday, usually invisible, violence.[52] As such, their deaths were re-experienced on a daily basis by family members confronted with news stories of the bombings, and then during the trials of the perpetrators in the case of Oklahoma City and the nearly daily invocations of 9/11 in the mass media and by politicians during the decade after the event. The publicity associated with these deaths created an extended bereaved community, and hence a community of mourners that both comforted and intruded into the lives of the families of the dead. Thus, the identity of the victims "not only signifies the relationship between a name and asset of physical remains but also encompasses the social ties that bind a person to a place, a time, and, most importantly, to other human beings."[53]

The creation of this community of mourners highlights the three different ways that victims of mass atrocity become recognized: first and most obviously, scientifically through the actions of forensic science and DNA identification; second, socially, when family, friends, and communities accept the scientific claim of identity made by the authorities; and finally, collectively, when the person is recognized by the broader national and international community through public commemorations and memorials, and fitted into a historical narrative about the past.[54] There is, of course, spatial and temporal separation between each of the three moments of recognition, but they are interwoven in ways that will be explored throughout this book.

Memorials and the Nation

Memorials to tragic events are expected to serve many purposes: to remember the event; to explain its historical importance; to mourn the deceased; to remember their lives and give their deaths broader meaning; to highlight the threat of terrorism and violence; to tell the perpetrators and the world that the goodness of humanity was not defeated by terrorism or violence; to serve the local community through nature, the arts, and public space; and to serve the nation as a source of pride, resilience, and political meaning.[55] Memorials fundamentally remember events that were unexpected and create a disjuncture in our understanding of history and ourselves.[56] They also suggest that the future should be different in some way. The decision to memorialize a set of past events plays a part in narrating what type of future should exist, both at the site of the memorial and in the society doing the remembering.

When an event is memorialized, there is a certain appeal in focusing primarily on the redemptive narrative identified by Linenthal in the aftermath of the Oklahoma City bombings—showcasing the better side of human nature, the pride and work ethic of a city or region, the essential goodness of the nation and its people, and the capacity for regeneration and redevelopment in areas affected by terrible events. Perhaps because of the desire for redemption, since World War II, and especially since the 1980s, American memorial culture has been expanded and democratized, both in the sense of which events get memorialized and who has a say in how the memorialization will be done. The time that passes from event to memorial has also been compressed.[57] Rather than discussions about memorialization taking place years or even decades after something happened, they often start days after an event.[58] There is an intense desire not to forget tragedies and lives lost, and to convey some sort of message—especially a positive one—to future generations about the event.[59]

The recovery, identification, and memorialization of the victims of the September 11 World Trade Center attacks brought science to bear on questions of identity, politics, and memory. The promise of identifying human remains through genetic technologies has fundamentally altered the way we will memorialize the dead in the future. These changes are not unambiguously good. Will the value of victims be measured by how much technology is applied to identify them? Will the loss of memorials to the unknown dead, and the lack of a clear end to identification efforts, change the way we remember traumatic events—both individually and as communities? How will the culture of memorialization change when there are no longer anonymous remains that belong to the collective, but only remains awaiting ever more powerful technologies to be identified and repatriated to families? These questions cannot be answered yet, but the response to the September 11 World Trade Center attacks can at least provide us with some clues. It is to this story that we turn now.

1

A Tuesday Morning in September

Neither words nor images can adequately describe the physical chaos that enveloped the World Trade Center site on the morning of September 11, 2001 and in the months that followed. Photographs and video can of course capture the sheer destruction that occurred as the twin towers collapsed floor by floor, the tons of debris reaching speeds of 120 miles per hour as they neared the ground. First-person accounts preserve the visceral fear that people in lower Manhattan felt as they either sought to escape the World Trade Center, or to rush in to try to save lives. Artifacts preserved in museums and archives can provide a reminder of the power of the falling buildings and of just how little actually survived the collapse. And missing persons posters and spontaneous memorials remind us of the lives that were lost that day.

Yet, when prompted, most people who were at the site on the morning of September 11 say that the most vivid memories they have are aural: the initial sound of airplanes flying far too close to the ground, the thuds that accompanied the impacts of the planes into the towers, the thunderous rumble of the towers collapsing, and the deafening silence that enveloped the area in the time between the collapse and the start of the frantic rescue efforts. Survivors, first responders, and bystanders also point to the smell that enveloped the WTC; one that seemed to be equal parts burning wire, concrete dust, jet fuel, and death.

The details of the attacks are well-known and easily accessible. Video of the event both from the media and eyewitnesses can be found all over the Internet—on YouTube, news and government websites, and on countless conspiracy theory web pages. Given the extraordinary global interest in

9/11, hundreds, perhaps thousands, of written accounts exist, each with particular strengths and weaknesses, differing degrees of accuracy, and radically different perspectives on the truth of the standard story of the day as told by government officials and the mass media. Readers wishing for detailed reconstruction of the events should start with NIST's *Final Report on the Collapse of the World Trade Center Towers* (2005). Jim Dwyer and Kevin Flynn's *102 Minutes: The Untold Story of the Fight to Survive Inside the Twin Towers* (2004) is an excellent description of what it was like for the people struggling to evacuate the towers after the attacks. Dennis Smith's *Report from Ground Zero* (2003) is a good account of 9/11 from the New York City Fire Department's (FDNY) perspective. A variety of firsthand narratives of the attacks and their consequences are collected in Damon DiMarco's *Tower Stories: An Oral History of 9/11* (2007).

Reconstructing the Events of 9/11

At 8:46 am, five hijackers flew an American Airlines Boeing 767 into the north face of WTC Tower 1 (the "North Tower").[1] Flight AA11 had recently departed from Boston's Logan Airport and was bound for Los Angeles with seventy-six passengers and eleven crewmembers. The plane was estimated to weigh approximately 283,600 pounds on the day of the attack. It struck the building at a twenty-five degree angle at approximately 440 mph, creating a large gash in the exterior skin of the building between the 93rd and 99th floors. The North Tower suffered significant damage to the external skeleton and interior structural core, but withstood the impact quite well. Perhaps the most significant effect of the crash, according to most engineering accounts, was that the impact knocked off the poorly installed insulation that was supposed to protect the steel columns at the core of the building, and the floor support system, from fire. This turned out to be a critical problem. The remaining undamaged steel in the building began to weaken when the jet fuel, and then office furniture, paper, and other contents of the building, burned.[2]

The first firefighters arrived on the scene four minutes after the impact, but they were already powerless to defeat the flames and smoke in the tower. They could not carry much of their equipment up to the fire because the elevators were out, and the building's fire suppression system had been destroyed in the initial impact. At 8:52 am, the first person was seen falling from the building.

By 9:00 am, FDNY chiefs had set up an incident command station in the lobby of the North Tower. They quickly began sending firefighters to the higher floors to evacuate people from below the impact zone and to reach as many survivors as possible above the impact zone. From analysis of communications and subsequent interviews, it is clear that FDNY personnel believed that a partial collapse near the top of the tower was possible, but they did not think the entire building would fall. FDNY officials signaled a fifth alarm, asking for additional resources to be staged at the corner of West and Vesey streets.[3]

At 9:02 am, as the North Tower was being evacuated, United Airlines Flight 175, also a Boeing 767 heading from Boston to Los Angeles, crashed into the south face of WTC Tower 2 (the "South Tower"). UA175 weighed less, but was traveling 540 mph, about 100 mph faster than AA11. It hit the South Tower at an angle of 38 degrees between floors 77 and 85. The angle and velocity at which the plane hit the building caused the floors above the impact zone to twist and rotate counter-clockwise. The South Tower's structure sustained more initial damage than the North Tower, and was weaker as a result. Just as in the North Tower, the plane's impact stripped core columns and supporting steel beams of insulation, making them vulnerable to the heat of the burning jet fuel and the contents of the plane and offices.[4]

FDNY almost immediately issued a second fifth alarm call (this time for the South Tower) and asked for resources to be staged at the corner of West and Albany streets. At 9:47 am, a third fifth alarm was issued, with orders for units to come to West and Vesey for instructions.[5] By this time, communication was starting to break down within the FDNY structure—too many people were using the wireless radio system, which did not perform well in high-rise environments. The communication system within the towers themselves had been destroyed by the impact of the planes and subsequent fires. Partly as a result of these communication problems, many of the units responding to the fifth alarm calls went straight into the towers without checking in or receiving formal instructions. They recognized the severity of the situation and wanted to help save as many lives as possible.[6]

Phone calls received by relatives and friends of surviving occupants in the two towers between the impact of planes and the collapse of the towers (many of which have been made public), along with calls to 911 (parts of which have been released by the City of New York), suggest that the situation very quickly deteriorated from confusion but relative calm to sheer terror and knowledge of imminent death. While the first calls generally alerted recipients that something had happened at the World Trade Center, callers did not seem particularly scared and stated that they were awaiting instructions on what to do. Later calls, however, included professions of love and acknowledgement that the caller and recipient would likely not see one another again. Sadly, many calls to friends and family went straight to voice mail and were stored by phone companies for later delivery because the damaged phone systems quickly became overburdened by the volume of calls being made. Many recipients did not receive these messages until a few days after 9/11. As a result, frantic efforts were made to rescue people who were thought to still be alive before the sad realization that these messages were the final recordings of victims who had already died.

At 9:58 am, the section of the South Tower above the impact zone began to lean to the southeast, and it quickly buckled. Over the next several seconds, the building collapsed floor by floor, as the material from higher floors smashed through lower floors like a group of dominos falling. By 9:59 am, the remains of The South Tower were spreading through lower Manhattan in a concussive wave.[7]

At 10:00 am, sensing for the first time that the North Tower was in imminent danger of collapse, the New York City Police Department (NYPD) and FDNY ordered all emergency responders to evacuate. While most NYPD officers heard the call, the majority of FDNY personnel—particularly those on the higher floors—did not receive it due to the problems with their communication system. Some who did hear the order may have ignored it in an effort to save more people. At 10:06 am, an NYPD helicopter radioed to police commanders that the building was breaking down at the impact zone and would likely collapse, and at 10:21 police helicopter personnel reported that the building was starting to lean to the southwest. FDNY firefighters still in the North Tower almost certainly did not obtain this information. At 10:28 am, the North Tower collapsed in a twelve-second free fall.[8]

NYPD and FDNY officials did not have strong working relationships and it was inconceivable for them to establish joint incident command operations, or even station personnel from each organization together to exchange and relay information.[9] Mayor Giuliani had created the Office of Emergency Management in 1996 to address this problem, but OEM had no success in bringing the two departments together.[10] NYPD personnel had much more information about the condition of the towers, and were able to communicate effectively in high-rise environments. This fact at least partially accounts for why only 23 of their officers died, compared to 343 firefighters. The Port Authority Police Department (PAPD) lost 37 officers that day.[11]

Amazingly, of the approximately 7,545 people who were in the North Tower below the impact zone, all but 107 safely evacuated thanks to a combination of instinct, emergency preparedness, and the bravery of rescue personnel. For those above the impact zone, the story was the opposite. All 1,355 people above the 92nd floor died on September 11. Many died as a result of the initial impact and fire, and the rest died because all means of escape had been cut off by the plane's impact. All stairwells and elevator access had been destroyed, and the NYPD determined that a roof rescue was impossible due to intense smoke and heat and the difficulty of landing on WTC roofs in the best of circumstances.

There were two major differences between the situations in the towers that led to a greater overall survival rate in the South Tower compared to the North Tower. First, by 9:02 am, many of the 8,600 occupants of the South Tower were aware that something horrible had happened in the North Tower and had already begun to evacuate—despite public address announcements to the contrary. Second, unlike in the North Tower, one of the three stairwells remained passable, providing an escape route for those above the impact zone. As a result, only 21 percent—619 of 2,900—of those individuals at or above the impact zone of the South Tower died when the buildings collapsed. Only 0.2 percent—11 of 5,700—of people below the impact zone died.[12]

On the Outside

Eyewitnesses reported steel beams weighing thousands of pounds flying through the sky like toothpicks, plane parts scattered about, pulverized concrete raining down on the streets below, paper floating from the twin towers like confetti, and, horrifically, bodies and body parts falling from the sky above. NYPD inspector James Luongo, who went on to be the police department's incident commander of the remains recovery effort at Fresh Kills Landfill, described it in a 2002 interview:

> We [a group of NYPD personnel who were walking on foot to respond to the emergency before the first tower fell] started working our way down Vesey St., which was dangerous. The debris was falling . . . and

it was total chaos. . . . We were trying to time the debris as it was coming down cause it was landing around us, and then we were noticing that it wasn't only debris coming down but bodies. So we ran down Vesey St, timing the bodies that were coming down, and a few of them landed around us, while we were hugging the wall as much as we possibly could. At one point in time we had misjudged one of the bodies: we didn't see one of the bodies coming down. So we had to hide underneath the pedestrian overpass that crossed Vesey from building six to building seven. I remember putting my face into the wall, my hands over my hat because I just didn't want to get sprayed with the human remains when the body hit. I just didn't want to get sprayed in the front, I didn't want to walk around like that all day. In the back, at least I wouldn't have to look at it, you know? Fortunately I wasn't sprayed though. When the bodies hit the ground, they were disintegrated. They were just splattered, like watermelons. It wasn't what I was used to. I'd dealt with jumpers before in my twenty-two years but I wasn't used to this. And there was a piece of the plane landing gear in the middle of the street, there was a huge spring from the airplane in the middle of the street, there were legs strewn apart, different pieces of human remains, across Vesey St.

There was a certain point where you could not go any closer, and if you went any closer you would get killed: the debris and bodies were raining down you just could not go past that. At one point in time on West St., I saw a woman . . . as she was getting her way out of the plaza area. She just disappeared. The debris came down . . . gone. It squashed her, that was the end of her and I didn't see any more of her, just a piece of debris fell down on her, and she was gone. But you could not get into that area. . . . We evacuated the area as best as we could. Total chaos, people running, people screaming. I stopped looking [at] the bodies that were coming down. I was in an area where I wasn't gonna get hit with the bodies anymore but you knew when the bodies were coming down because the people around were screaming as the bodies were coming down. And uh, I just stopped. You know you say a prayer, what are you gonna do, you know?[13]

The dozens of other eyewitness accounts from first responders (made public through a FOIA request by the *New York Times* and a group of victims' families) and survivors describe a scene similar to Luongo's. Tom

Haddad, who survived the attacks in his office on the 89th floor of the North Tower, recalled that, as he was evacuating the building, "occasionally you'd hear these devastatingly loud thumps. At the time, I thought they came from more falling pieces of the building. It didn't register, there were hunks and piles of meat all over the ground . . . nothing I recognized as body parts. Later on, I found out they were the remains of jumpers."[14] Jules Naudet and his brother Gedeon happened to be filming a documentary about rookie FDNY firefighters in lower Manhattan that morning and followed their subjects to the site; these thumps can be heard on their footage as they approached the burning towers.

There is no good estimate of the number of people who jumped or fell to escape the hellish conditions at the top of the towers or hasten inevitable death. While NIST's report on the collapse of the towers puts the number at 111 for the North Tower, other sources suggest that up to 200 people jumped or fell from the towers.[15]

Whatever the number, images of falling victims have remained a raw and chilling part of the visual record of 9/11. One image in particular, dubbed "Falling Man," by AP photographer Richard Drew, has inspired tremendous debate. It was published in newspapers around the world in the days after 9/11, but most media outlets refused to show it again after that.[16] It inspired a documentary (*9/11: The Falling Man*), plays a significant role in Jonathan Safran Foer's 2005 novel *Extremely Loud and Incredibly Close,* and serves as a central element of Don DeLillo's 2007 novel *Falling Man.*

The collapse of the towers is devastating when viewed on video. First-person stories of survival during the collapse bring home the terror that people in the shadows of the towers must have felt as they ran for their lives. Forensic anthropologist Amy Mundorff had just arrived at the World Trade Center with a team of colleagues from the New York City Office of the Chief Medical Examiner (OCME) to evaluate the scene and get a sense of the number of bodies that would be arriving at the morgue. In 2012, Mundorff recounted her experience:

> So four of us from that team went down in one car and Dr. [Charles] Hirsch and a medical legal investigator and one of our operations guys and the driver went down in his car. We were all going to meet down there. And we parked on Vesey Street and there were things, building parts and people were jumping. We walked over . . . across from the Marriott Hotel and we were deciding something but I cannot, for the

> life of me, remember what it is anymore. But we were going to . . . someone was going to go back and get a camera and we were going to regroup there. Then I heard this rumble that was loud and my [colleague] said, "Oh, a building is coming down." And I turned around and just started to run. And I could see in front of me the stairs for World Financial One, so we were in the kill zone. And that tidal wave just picked me up and blew me into a wall. I mean, it was so powerful where we were that the building I was in front of, the stone façade was torn off. It was so powerful. . . . So when it all stopped, you could just hear beeping and it was like the firemen's alarms and car alarms and it was pitch black. I knew I was in front of the stairs, so I felt my way up the stairs to get to the building. When I got to the building door, I came across one of my colleagues. And when we got inside the building, we found another one. So one of them we could not find. And it was just like chaos and panic.[17]

Mundorff credits her mountain-climbing husband for giving her the advice that may well have saved her life: in an avalanche, make sure you cover your head with your jacket and provide yourself with an air pocket in case you get buried. On instinct, as the debris cloud enveloped her and threw her into an exterior wall of a building, she remembered his advice and managed to survive. Eventually, she was able to find two of her colleagues—one of whom had received a severe head injury and sustained brain damage—and they made their way to the water's edge behind One Financial Center, where a police boat took them to Jersey City. After receiving stitches for her injuries and finding out she had a broken rib and a concussion, Mundorff spent a sleepless night and a restless day at her parents' home in Armonk, New York. She returned to work on September 13, knowing that her anthropological knowledge would be needed as victims started arriving at the morgue.[18]

In interview after interview, eyewitnesses describe the force of the collapse and subsequent debris cloud as something incredibly powerful. In an interview in 2002, firefighter Timothy Burke described the cloud as a "three-dimensional object" that you did not just breathe in, but ate.[19] The collapses forced the pulverized material deep into the lungs of those who were in the vicinity when the collapses occurred, and caused people to vomit up the dust while simultaneously struggling to gasp in enough oxygen to survive. The cloud ultimately blanketed much of lower Manhattan in a thick layer of a grayish powder that consisted of concrete dust, the crushed contents of two of the largest office buildings in the world, and human remains.

After the Collapse

People who arrived on the scene just after the towers collapsed report being awestruck by the devastation with which they were confronted. Life-long New Yorkers found themselves completely disoriented on streets that they had walked innumerable times before. Others felt like they were in a movie, or had stumbled into a warzone. "Surreal" was a common descriptor.[20]

For journalist William Langewiesche, whose 2003 book *American Ground: Unbuilding the World Trade Center* (first serialized in *The Atlantic* in 2002) was criticized for its unvarnished portrayals of the people involved in the initial response to the attacks and the cleanup of the site, the reaction was different. "After years of traveling through the back corners of the world, I had an unexpected sense [of] . . . familiarity. Wading through the debris on the streets, climbing through the newly torn landscapes, breathing in a mixture of smoke and dust, it was as if I had wandered again into the special havoc that failing societies tend to visit upon themselves. This time they had visited it upon us."[21]

Before the dust had even begun to settle, firefighters, police officers, paramedics, federal officials, construction workers, and ordinary citizens returned to the site to search for survivors. Hospitals in the area mobilized to treat the rush of traumatic injuries that they assumed survivors of the collapse would have. People who knew their loved ones were in or near the twin towers hurried to the scene, looking for their spouses, children, parents, relatives, and friends. Although some of the searches were successful, thousands of families had no luck. Verizon, the dominant telecommunications company in the region, worked to reestablish contact with any trapped survivors who may have had functional cell phones.

The initial effort to locate survivors from the collapse of the towers was barely controlled chaos. First responders and volunteers alike searched the twisted and contorted rubble for signs of life, not always with regard for the structural integrity of the spaces that they explored.[22] Relying on instinct and adrenaline rather than training, they formed bucket brigades and commandeered any light construction equipment that they could find in order to search pockets of air amid the wreckage. Complicating matters, there was no effective credentialing system at the site until September 16, so large numbers of unnecessary personnel became involved in the search and rescue effort, making it difficult to work efficiently and potentially putting peoples' lives at risk.[23]

Of all of the groups that arrived on the scene, the stories of fathers searching for their sons were among the most poignant. Because firefighting tends to run in families, many of the firefighters who went missing on the morning of September 11 had other family members in the FDNY or other responding fire departments. They continued the search for months after the towers collapsed. Many of them became known as the "Band of Dads" and were regularly featured in media reports and a well-regarded photography collection by Gary Suson.[24]

Care was taken not to further damage any potential spaces where survivors might be trapped, but few people at the site were truly hopeful that anybody could survive the hellish conditions below ground. After all, in addition to the collapsed buildings, potential survivors had to contend with water both from heavy rain and efforts to put out the intense fires that burned deep within the pile.

Sam Melisi, a member of Rescue 3, FDNY's Building Collapse Unit in the Bronx, acted as a liaison between the firefighters and the construction and engineering personnel. In an interview from 2002, he recalled thinking optimistically that "since these were some of the world's largest buildings . . . we were going to find some of the world's largest voids. I was very hopeful that we were going to start finding people right away. It didn't pan out all that well."[25]

Hoping that their loved ones had been injured in the collapse and taken to nearby hospitals, those searching for the missing called or visited the major trauma centers regularly in the days after the attacks. They also posted missing persons notices wherever they might be seen by someone with news of their loved ones' whereabouts.[26]

At this point, nobody had a clear sense of how many people were actually missing. In order to collect missing persons reports and gather any antemortem data that might help investigators identify victims, NYPD and the Red Cross, along with other organizations, established the Family Assistance Center at the NYU medical school in the hours after the attack. This facility was soon moved to the Lexington Avenue National Guard Armory, and then to Pier 94, where it remained open until the end of 2001 when it moved to Chambers Street. Family Assistance Centers (FACs) were also opened in several other locations in the New York metropolitan area, including Queens, Staten Island, Long Island, and Liberty Park in New Jersey.

Robert N. Munson, who was emergency service director of the Minneapolis Area Red Cross, worked at the New York City FAC from September 23 to October 14, 2001. He described it as

> a giant pier warehouse building—set up like a trade show with full carpet, poles with drapes, and some 75 agencies with all their staff and stuff; service agencies, government, immigration, FBI, child care, etc. It is a comfortable building in order to be welcoming and calming as can be to the families. The City of New York has done a good job. It is a full city of people in this building—with free meals for families and separate 3 meals a day for workers. The dining areas are nice (as they can be)—donated fresh flowers daily, tablecloths, and an ambiance of peace and calm different from the noisy, bustling activity everywhere else. There are clients and workers everywhere. Lots of noise. People come here to access a broad array of services. Security is the tightest I have ever experienced anywhere. You can't get near the place without going through several barricades, body and bag checks—and once in, all workers need to be separately badged daily even though we have permanent clearance badges. Clients, of course have ID, are escorted, and limited by lots of armed police and military to certain areas. Interestingly, several dogs around with handlers—not sure if they are rescue dogs off duty or other "sniffers" and working dogs. Outside is where the wall of photos is that you see on TV—it is so moving that I avoid spending too much time passing by.[27]

The FACs were designed to be one-stop-shopping for relatives of the victims and others directly affected by the attacks, such as residents who were unable to return to their homes or people who had lost their jobs as a result of the attacks. Visitors could receive mental health services, apply for financial assistance, food stamps, welfare, housing assistance, and other social services, get help with life insurance, and receive free legal advice. Those who were missing a loved one were also asked to provide information about the person and a sample of their own DNA as well as any items—toothbrushes, hairbrushes, or razors—that might contain DNA from the missing person.

Although it was not known at the time, the NYPD's lack of experience gathering information and biological samples on such a large scale caused several problems.[28] Most notably, the police allowed anyone to fill out a missing persons report, regardless of their relationship to the person in question, and did not accurately record the biological relationship of the person who provided a DNA sample. This led to multiple missing persons reports for the same person, but with variations in basic physical descriptors like height, weight, distinguishing features, prior injuries, and clothing worn as well as slight differences in the spelling of names and the reporting of birthdates.[29]

These errors made it very difficult for the OCME to use the information and biological samples for matching and necessitated a resampling effort several months later. Further, given that so many of the victims of the attacks lived outside the New York metropolitan area, the FAC in Manhattan could only deal with a fraction of the need for such facilities. According to Robert Shaler, the OCME had hoped to play a leading role in the collection of antemortem data and DNA samples, given their responsibility for identifying the missing and dead, but NYPD would not allow them to take on this job.[30] In OCME medicolegal investigator Shiya Ribowsky's view, "the decree from City Hall that placed the cops in charge of the efforts to obtain DNA from victim families had turned into a debacle," and the firefighters' efforts to collect samples from their families was little better.[31] In addition to clerical errors and missed opportunities for collecting direct samples (such as biological samples from recent medical procedures), police and firefighters also lacked a basic knowledge of genetics and often failed to collect appropriate familial samples.

The Pile

Back at the pile, the grim reality of September 11 was becoming apparent to all involved in the rescue efforts. After a few dramatic rescues, there was nobody left to save. Anyone still under the rubble was dead, not trapped. And it was already becoming clear that most victims' bodies were destroyed beyond recognition by the impact of the planes or collapse of the towers. *New York Times* reporter Dan Barry informed readers that, by the late afternoon of Wednesday, September 12

> the jaws of huge cranes were biting indiscriminately into the piles of rubble, while police officers, firefighters, soldiers and other rescue workers pried at the ground with shovels and crow bars to free body parts, bits of human flesh, and rubbery patches of skin. Then, like

> sanitation workers tending to some hellish park, they carefully dumped the scraps of human remains into a green trash bag held open by a soldier. At times, men gathered to puzzle over a piece of flesh on the ground; dogs sniffed at the bits with little enthusiasm and moved on. 'We don't find much,' said a firefighter from East Rutherford, N.J.[32]

The only consolation that the families might receive was that subsequent forensic examination of the remains revealed little posttraumatic response (e.g., swelling or blistering), meaning that the majority of victims died soon after they were injured.[33]

Although the search and rescue effort officially continued for nearly a month, just a few days after the attacks, Mayor Giuliani was already gently informing New Yorkers and the rest of the world that there was little chance of locating more survivors. "The recovery effort continues and the hope is still there that we might be able to save some lives," he said. "But the reality is that in the last several days we haven't found anyone."[34]

Although Giuliani was forced to deliver this bad news to his city and the world, he leavened it with a promise that would ultimately lead to the largest identification effort ever undertaken. He mandated that the OCME do whatever it took to identify the source of every single human remain recovered from the WTC site, no matter how small.[35] This meant that unlike most disasters, in which investigators take an "identify all victims" approach (that is, make sure to identify at least one part of every person thought to be involved in the incident to confirm their presence), the OCME had to keep the identification efforts moving forward until there were no remains left to be identified or they had exhausted the limits of technology. The OCME has taken this mandate seriously and continues to push the limits of DNA identification techniques in order to meet it. This effort, however, it has taken a significant emotional and resource toll on the organization.[36]

A New York Effort

The Federal Emergency Management Agency (FEMA) immediately activated several of its on-call urban search and rescue teams to assist local rescue workers at the World Trade Center site, but neither that agency nor the Army Corps of Engineers nationalized the site. Firefighters became the de facto leaders of the effort to recover the remains of those who did not make it out before the towers collapsed.

Initially, first responders believed that living people were trapped in the rubble and didn't want to wait for equipment to be trucked in from far away to search for them. Sam Melisi said:

> These were people you had worked with, and they were maybe alive. You knew they were trapped in there, and there was a sense of franticness, and it was personal. I remember crawling through the steel—it would have probably been by the hotel. There were some spaces that let you get below and take a look around. It wasn't regulated at all. The first couple of days, anything went. It wasn't like somebody was saying, 'You can't go in there, you can't do this, you can't do that.' It was more, 'Hey, if you think you can get in there, go ahead.' All bets were off. It was just 'Go and bring somebody home.'[37]

Salvatore Torcivia, a firefighter who had been a police officer before joining FDNY, confirms the free-for-all nature of the initial response. The area was so large that there was little coordination of the search field—one simply chose a section of real estate and dug in. Once the search was complete, the worker would use a can of spray paint to mark the area.[38] Looking back on the recovery efforts a year after 9/11, Richard Garlock, an engineer who advised the city on structural issues at the WTC, said he was often astonished to see the spaces that had been explored by firefighters and others during the hectic first hours of the search effort.[39]

The Struggle for Control

After a few days, order slowly emerged at the site. Like everything else that would happen at Ground Zero in the years after 9/11, however, this process was riven with political infighting and media-fueled controversy. In the wake of the 1993 World Trade Center bombing, New York City officials knew their city was a target for terrorist plots and made some efforts to prepare. One immediate problem city leaders faced was that the departments devoted to public safety, the FDNY and the NYPD, had been intense rivals for decades. In an era of declining budgets and reduced workforces, the departments fought for resources, jobs, and prestige. In his vivid and moving book *Closure: The Untold Story of the Ground Zero Recovery Mission,* Port Authority Police Department Lieutenant William Keegan Jr. explains that "except in their most basic roles—law enforcement and firefighting—each was intent on outdoing the other. The police department had a scuba team; so did the fire department. The fire department had rescue units; so did the police department."[40]

More problematically, each department operated autonomously and was not inclined to share information or establish joint command in the event of a large-scale event. The Office of Emergency Management was supposed to solve this problem. Keegan believed that the city would be much safer and more efficient if the departments worked together to respond to emergencies. Unfortunately, because of a combination of bad planning and bad luck, the OEM's command center was located on the 27th floor of WTC 7 and was destroyed when that building collapsed on the evening of September 11. Thus, it was unable to take the lead in planning the response to the attacks.[41]

In this power vacuum, the NYPD claimed immediate control over the site because it was considered an active crime scene. But once the District Attorney decided to close its investigation of the attacks, which it did very quickly because there was no doubt about what had happened to the World Trade Center, the FDNY took over.[42] With 343 firefighters missing, it felt an obligation to find them. Technically, the FDNY had an advantage over the NYPD: it had authority over collapsed buildings and all buildings on fire. The FDNY ended up having sole control of the site through the end of October, when they were forced to share (or more technically, cede) power to the city's Department of Design and Construction (DDC).[43]

In his book, Keegan describes the initial frustration that he and other Port Authority Police Department (PAPD) officials had when dealing with the much larger NYPD and FDNY. PAPD officials felt they had been left out of the loop in initial rescue efforts despite having suffered a significant loss of life—37 out of a force of 1,100—and their familiarity with the buildings. Because the PAPD was not part of the city, there were no preexisting lines of communication between it and either of the other uniformed service agencies. Neither the NYPD nor the FDNY made any effort to inform the PAPD about what was going on, or to ask for assistance. Keegan also felt that the competitive atmosphere that pervaded the relationship between the police department and fire department meant that neither had much interest in dealing with the much smaller and less powerful PAPD.[44]

While the FDNY assumed authority over the search and rescue effort, the little-known DDC took charge of mobilizing the construction community to assist the firefighters and to begin the process of cleaning up the site. The DDC was largely responsible for managing all public construction projects in the city, from sewers and roads to firehouses and libraries. The DDC's two top officials, Kenneth Holden and Michael Burton, believed they had all of the necessary equipment available in the area and were prepared to handle the job.[45] Although the federal government would pick up the tab for the cleanup, New Yorkers would ultimately do most of the work. Within a few days, four major construction contractors had been hired by the DDC: AMEC Construction Management, Bovis Lend Lease, Tully Construction, and Turner Construction. Holden and Burton divided the area into quadrants and gave each company responsibility for removing the debris from one of them.

They faced a significant challenge: dynamite and other explosives that were normally employed by the demolition industry could not be used because of the instability and complexity of the site. Debris had to be hand cut and hauled out one piece at time.[46] "The pile was an extreme in itself," wrote Langewiesche continuing:

> It was not just the ruins of seven big buildings but a terrain of tangled steel on an unimaginable scale, with mountainous slopes breaking smoke and flame, roamed by diesel dinosaurs and filled with the human dead. The pile heaved and groaned and constantly changed, and was capable at any moment of killing again. People did not merely work to clear it out but went there day and night to fling themselves against it. The pile was the enemy, the objective, the obsession, the hard-won ground.[47]

By the end of the first week, the basic structure was in place: FDNY was in charge of the rescue and recovery effort and the four New York construction companies were in charge of removing debris from the site. As the effort wore on, this arrangement would lead to significant strife and controversy. As Keegan notes, "From that moment on, there would be those of us who saw Ground Zero as a rescue and recovery site and our job as being to find fallen heroes for their loved ones, and those who saw it as a construction site and their job as being just to clean it up."[48]

Firefighters directed the effort to recover human remains and had complete authority to tell construction workers where to dig, at least through the end of October. They worked in teams of seventy-five firemen, supplemented by NYPD and Port Authority police officers. The search effort was a three-part process. Initially, uniformed service members stood by, armed with rakes and shovels to examine newly loosened debris, as construction workers pulled the steel and rubble piece by piece from the site. If suspected human remains were discovered, these were removed and brought by FDNY Emergency Medical Services escorts to one of the temporary morgues set up around the site. According to at least one account, every remain recovered was prayed over before being removed.[49] A second inspection was done at the debris-transfer point before it was loaded onto trucks and barges to go to Fresh Kills Landfill for processing and disposal, where a third inspection was performed to locate anything that may have been missed on site.[50]

The firefighters initially concentrated their efforts on the places where they expected their own to be found, but the site never yielded the "mother lode" of bodies that everybody was hoping for. As the recovery effort got underway, the construction companies—especially Bovis Lend Lease—played a supporting role. The uniformed services told Bovis superintendent Charlie Vitchers and his management team where they wanted to search, Vitchers consulted structural engineers to make sure the area was relatively safe, and, if so, cranes and other machinery were brought to the area to assist in the removal of debris.[51]

Search personnel recovered body parts of varying sizes and degrees of recognizability. In the first few days, body parts were still relatively intact and easy to spot, but as they begin to decay, they started to look like the rubble and debris in which they were supposed to be found. After a few days, the best indicator of remains was the smell. Torcivia recalled that "the smell got to everyone after a while. By the second or third day, it was raw down there. I was gagging. If we uncovered anything that resembled a body or a body part, we'd shovel it into a body bag—there was no other way to pick up what we found. We hoped they'd be able to ID the remains through DNA testing down at the medical examiner's office."[52] More complete bodies were rare. For the most part, the remains of firemen were the best preserved because of their heavy bunker gear, while the remains of women were the least well preserved because they tended to wear the lightest clothes.

Firefighters—both living and dead—quickly emerged in the press and popular culture as the heroes of the tragic situation. They were described as selfless patriots who marched into battle, in sharp contrast to the cowardly and barbaric terrorists who had murdered thousands of innocents. Although most firefighters resisted the urge to play up their heroism, there was at least some resentment among the police and construction workers.[53] In Langewiesche's account, nowhere were these tensions more pronounced than in the treatment of human remains: "On the one extreme the elaborate flag-draping ceremonials that the firemen accorded to their own dead, and on the other, the jaded 'bag 'em and tag 'em' approach that they took to civilians."[54]

Firefighters offered differential treatment to their own victims compared to other uniformed service victims and civilians. Each time the remains of a firefighter or firefighting gear was located, the entire Ground Zero recovery effort ceased. All workers at the site, whether uniformed service personnel or civilian construction workers, paused and saluted the fallen firefighter. If remains were recovered, they were wrapped in an American flag and paraded through a lineup of Ground Zero workers.[55] These remains were immediately sent to the OCME via ambulance rather than being placed in a refrigerated trailer on site for later transport to the OCME.[56]

More pernicious than the problems that emerged as byproducts of the need for speed were the overt efforts of FDNY recovery personnel to "reconstruct" the remains of their fallen brethren—by placing body parts found adjacent to empty or partially empty articles of clothing into that clothing—in order to be able to provide their loved ones with as complete a body as possible. This manipulation was seen only with FDNY clothing, and was usually obvious upon examination—for example, when parts of a leg were stuffed into the sleeves of bunker gear or when boots that were associated with a set of bunker gear contained two left feet.[57] FDNY personnel often removed personal effects, especially jewelry, from their brethren to ensure that family members received them in a timely and comforting fashion. It is unknown whether such actions prevented the identification of any firefighters (as when the only recovered piece of an individual was placed with another body and not discovered through anthropological investigation or DNA identification). OCME personnel, however, did not confront FDNY personnel at the time because they recognized the grief that firefighters were dealing with and were not sure that their concerns would be taken constructively.[58]

Once human remains were found in the Pit, they were brought in bags to the body collection point (BCP), where medicolegal investigators briefly examined, catalogued, and prepared the remains for transport to the triage center at OCME headquarters on 30th Street between First Avenue and FDR Avenue in midtown Manhattan. OCME activity was limited at

Ground Zero—they were not included as part of recovery teams and were only occasionally consulted about best practices for removal of remains to ensure maximum evidence retention and minimal damage to physical integrity.[59]

It was often difficult for OCME personnel to know whether the location information provided for each remain was accurate or not. It was especially unclear how deep workers had dug to find a particular remain, which could help determine which floor the victim was on at the time of his or her death and speed the reassociation of remains. Since human remains were often discovered only after raw material had been removed from the Pit, OCME personnel also worried that the GPS reading (or grid coordinates) did not indicate the original location of the remain. Mundorff and others wanted the military's Joint POW-MIA Accounting Command to help set up the procedures for recovery of human remains, but the city never asked the agency for assistance.

Further, in the days after the 9/11 attacks, Sophia Perdikaris, an archaeologist at Brooklyn College, worked with the Society for American Archaeology to collect the names of 350 trained archaeologists who would be willing to support the recovery effort at Ground Zero. Neither the city nor the federal government asked for their assistance.[60] People familiar with the recovery effort say that FDNY could have benefited from more professional anthropological and archaeological expertise on site. After all, the FDNY did not have training in forensic archaeology, human anatomy, or human remains excavation. One major problem that resulted was the commingling of remains from more than one person in a single body bag, rather than the separation of remains that could not absolutely be related to one another into unique bags. Further, bodies in an advanced state of decomposition were not always handled properly during recovery.[61]

Nobody has suggested that FDNY botched a significant number of cases or that the recovery effort was marred by wholesale incompetence, but many believe an on-site anthropologist or archaeologist would have significantly reduced the number of commingled remains that arrived at the OCME facility, increased the number of more complete remains, and made some reassociations easier. This would have saved OCME personnel a great deal of time, effort, and stress. It also became an issue as families interacted with the OCME, desperate for information about where exactly the remains of their loved ones had been found. Many families wanted this information in order to piece together the last moments of the victim's life.

Where was he when he died? Who was she with? Had he tried to escape or did he stay put? Although information about the location of human remains was not usually reliable in reconstructing the last moments of a given individual, it did help families and friends make sense of the horrific event.[62]

In addition to using better recovery methods, professional archaeologists might have searched for human remains beyond the perimeter of the sixteen-acre World Trade Center site—which is where FDNY concentrated their efforts—even if it could not have been cordoned off by law enforcement. Richard Gould, now retired, was a professor of archaeology at Brown University at the time of the World Trade Center attacks, and a founder of what he calls "disaster archaeology."[63] Gould's wife was working in New York City at the United Nations and was in the city on September 11. Gould made his first visit to the site on October 6, 2001. In his 2007 book, Gould notes that what immediately struck him on this visit was the presence of large amounts of ashy dust and debris outside of the search and recovery area. Gould maintains that on this first visit, he saw what appeared to be small fragments of charred bone—the largest was about two or three inches across—lying in the dust. Although he had his camera with him, he says he did not attempt to photograph these remains or collect them for fear that he would alert the mourners and other visitors to the site of his discovery. On subsequent visits, Gould said he saw more of the same phenomenon: fragmented human remains mixed in with the ashy matrix that was expelled by the twin towers as they collapsed. He was surprised, however, that cleanup crews were making no effort to collect this potentially forensically relevant material—they were working as quickly as possible to clean up the thick dust that had settled on lower Manhattan, but did not have the remit to carefully examine what they were sweeping up and hosing down. Mayor Giuliani had ordered lower Manhattan to be cleaned up as quickly as possible, to restore a sense of normalcy and to remove potentially dangerous material from the streets and buildings around Ground Zero.

Gould's observations revealed both a profound problem and an opportunity to put to the test the principles he was developing in the field of disaster archaeology. On November 2, 2001, after communicating with officials from the Office of Emergency Management (OEM), OCME, NYPD, and FDNY, Gould and Perdikaris did an informal scan of twenty building roofs in the vicinity of the WTC site. They found that power-washing teams had already scrubbed the roofs, but that a significant amount of debris remained in airshafts and under ventilators.[64]

A few weeks after the search concluded, the *New York Daily News* published an article on the work of FDNY's Phoenix Unit to catalogue the locations of human remains using GPS technology.[65] An accompanying map showed that human remains were found where Gould had observed, but did not examine, human remains on October 6 and 7, 2001 and at the site of a subsequent March 2002 search he conducted with colleagues on Barclay Street. Although Gould could only speculate whether his reports led to the discovery of these human remains, it is clear that the "GPS locations provided independent confirmation that significant amounts of human remains were dispersed over wide areas of lower Manhattan outside Ground Zero. These reports also made the existence of such debris scatters public, precipitating a rush of inquiries by relatives of WTC victims to the FDNY and Medical Examiner's Office."[66] After the *Daily News* article was published, several other human remains deposits were discovered at other sites around lower Manhattan in late 2002 and early 2003. This issue would come back to haunt the recovery effort in 2005 and 2006, when more human remains were discovered in numerous locations in and around the World Trade Center site.

Sanctifying Remains

By early October 2001, city officials had come to the grim conclusion that many families of victims of the World Trade Center attacks were unlikely to receive remains of their loved ones. Rather than giving back nothing, however, the city came up with an innovative solution that would both show that it cared for the victims of the attacks and prevent families from purchasing World Trade Center debris from "profiteers."[67] If remains could not be located in the debris, then the debris would be transformed into a relic that could stand in for the bodies of the dead.

Mayor Giuliani ordered the NYPD to collect debris from the site, sanctify it through careful (though arbitrary and ad hoc) rituals, and then place a small amount of it in cherry wood urns to be delivered to families. In the first step of the process, powdered debris was shoveled into three fifty-five gallon drums at Ground Zero and blessed by a chaplain on site, then draped with American flags, transported with a police escort to One Police Plaza, blessed again, then guarded by two honor guards twenty-four hours a day in a room that was freshly cleaned, repainted specifically for this purpose, and fitted with potted plants to bring life to an operation that otherwise referenced only death. In the second stage of the process, the remains were carefully spooned into plastic bags, which were sealed and placed in high-quality wooden urns by gloved members of the NYPD's ceremonial unit. This act was done with great care, in a room with low lights and soothing music. The urns were then sealed, inspected, and stored for safekeeping until they were handed over to families in late October at a special ceremony at Pier 94.[68]

Trouble at Ground Zero

As the recovery effort wore on through September and into October, the emotion that fueled the frenzy at the site began to take its toll on everybody involved. Tensions among the various participants in the cleanup efforts were running high. Mayor Giuliani and his team felt that it was time for the city to move forward. There was no going back to September 10 to be sure, but the pace of the cleanup effort at Ground Zero had to increase. It was time to wrest control of the site from the firefighters.

Firefighters felt that they were waging a constant battle against political machinery that had a strong interest in cleaning up Ground Zero as quickly as possible. More than anything, they wanted to recover all of their dead, as well as the remains of civilian and other uniformed service personnel, and did not want political agendas or construction deadlines to prevent them from doing so. "They [the politicians] want to pull out people with cranes," one firefighter explained to the *New York Times*. "We want to bring back the brothers with dignity. They think the quicker they can clean it, the better they look. We've got friends there, brothers and family. We've known these guys for years. This goes very deep."[69] Construction workers often found themselves trapped in the middle of this battle—wanting to help the firefighters find as many remains as possible, but pushed by their bosses to get the job done on schedule.[70]

Tensions reached a boiling point on November 2. At a press conference, Giuliani made it plain that, throughout the recovery process, he never believed remains would be recovered for the majority of victims. "I've known from the beginning, from the first night, that it would be a burial ground. The medical examiner, the first night that I met with him on the evening of September 11, told me that the crush of the buildings and the high degree of heat was going to mean that . . . the majority or vast majority of people would disappear because they would evaporate."[71] Yet recognizing the importance of the recovery effort for the psyche of the FDNY, the city, and the world, he encouraged it and allowed it to proceed without much regulation.

The time had come, however, for a new order to be imposed. "The reality is, then," said Mr. Giuliani, "if you're the police commissioner or the fire commissioner or the mayor and you're responsible for the lives of other people, you have to say, well, maybe these people are too emotionally involved to be involved in this operation. Maybe these are not the people that have the ability to detach so they can handle it professionally."[72] The implication of this statement was not only that the role of firefighters ought to be reduced, but also that they had been making poor decisions on the job. Ultimately Giuliani used the issue of safety to scale back the number of firefighters engaged in the recovery effort—from between sixty-five and seventy-five to twenty-five—and to limit their ability to stop removal of debris from the site in order to recover human remains.[73]

Firefighters greeted the announcement with anger and indignation. Whatever the merits of the decision, its justification made little sense: in a round-the-clock recovery effort that had spanned more than six weeks, not a single firefighter had been killed or seriously hurt on the job. Several hundred of them converged on City Hall in protest, chanting "Bring Our Brothers Home." They also demanded the ouster of Giuliani and Fire Commissioner Thomas Von Essen, who they believed supported the mayor rather than the firefighting community. What started as a largely peaceful demonstration, however, quickly turned violent as firefighters clashed with police. Twelve firefighters were arrested and five police officers were hurt. The *New York Times* reported that the fight "rattled the city's top officials and laid bare the frustrations of the living who are unable to bury their dead."[74]

Despite efforts to put Giuliani's plan into effect, the firefighters ignored the new directive and continued to search for remains as aggressively as they could. In response, Giuliani spoke directly to the public about the firefighters' allegations. "They really are so off base that it's a sin," he said. "And I mean that in a moral sense. What they're doing is sinful. The effort here is to try to recover as many human remains as possible. They have absolutely no monopoly on caring about the people there."[75] He also made various conciliatory gestures—increasing the number of firefighters on search duty at the site to fifty and dropping charges against nearly all the protesters—but none mollified the firefighters.[76]

Family Activism

The search for, and treatment of, human remains motivated the families of victims into forming activist and lobbying groups. Although most of the relatives who became involved were not political activists before 9/11, they quickly found their voices when it came to the remains of their loved ones. Other than the male firefighters searching for their sons on the pile, the majority of these newly active individuals were women—mothers, wives, and sisters of the missing. Grief, marriage, motherhood, and widowhood became potent political weapons that politicians could not afford to ignore. Much like the women of the Plaza de Mayo in Argentina and Srebrenica in Bosnia before them, these activists used their special social status to demand that elected leaders do more on behalf of their loved ones.[77] A November 25, 2001 *New York Times* article highlighted this phenomenon: "A new political group has emerged from the cinders and ash of the World Trade Center disaster site. It is a group that no city official wants to offend, one whose broad powers are only beginning to be realized by its members. In the verbal shorthand of these troubled days, it is known simply as 'the widows'."[78]

One of the first activists to emerge was Marian Fontana, the wife of missing firefighter David Fontana and mother of a young son. She helped found the 9-11 Widows' and Victims' Family Association, with the support of FDNY union officials, to ensure that the needs of families and victims would not be trumped by the financial or political interests that had an enormous stake in cleaning up and redeveloping the site as quickly as possible. One of the first causes Fontana adopted was restoring the number of uniformed service personnel working in the Pit. Fontana used her identity as a firefighter's widow and the mother of a now fatherless child to support this cause. As news of the reductions in firefighters at the pile spread, Fontana asked to meet with Giuliani and Von Essen on November 9.[79] In an interview shortly after the meeting, Fontana told a *New York Times* reporter, "Yes, I realize the power we all have. I definitely want to do right by the families, foremost, and the firemen, and there is pressure to proceed. It's a little daunting sometimes."[80]

Civilian families were represented by the group Give Your Voice, which had recently been formed by the family of twenty-six-year-old apprentice electrician James Marcel Cartier, who died on the 105th floor of the South Tower, and others. Give Your Voice's mission was twofold: first, they wanted "to be sure that every effort is being made to spot, recover, preserve, identify and deliver any human remains to the families of the victims in order that a decent burial may be had by the families and loved ones." More generally, though, they wanted to provide a voice for civilian families that would complement that of uniformed service families, as well as a structure to funnel information from the city and other agencies involved in the recovery effort back to civilian families.[81] In their view, civilian families had been "forgotten" in the conversation and media coverage about 9/11 victims, and they wanted to remedy this travesty. The organization was not created in response to the activism of the uniformed services families, so much as in complement to it. In fact, Give Your Voice went out of its way to work hand in hand with those families. In their early web updates, they regularly thanked uniformed service families for their encouragement and support. Even their logo brought together the two stakeholders: one side of the image showed a man and a woman—a tradesman or construction worker and a woman in a suit, with a newspaper and briefcase—and on the other side was a firefighter's hardhat.[82]

Once the dignity of the recovery effort was ensured, Give Your Voice planned to turn to other issues, such as how to help families navigate the complex financial assistance system, in which dozens of organizations were ostensibly providing support to those who lost loved ones in the attacks. Indeed, in a November 17, 2001 family update, the organization noted that "to date there is no Unit designed to assist Families of Civilians other than the Pier, which has been a horrible experience for most Families. Nor is there any liaison in the Mayor's Office to offer information regarding any assistance. To date most of the information Civilian Families receive is from the Media and close Friends in the Fire Department."[83]

The November 9 meeting between Marian Fontana and Giuliani led to a larger meeting, on November 12, between the firefighters, the mayor, and senior officials associated with the World Trade Center response (coincidentally, the meeting was the same day that American Airlines 587 crashed in Queens as a result of pilot error and mechanical issues). The meeting lasted three hours. The families were represented primarily by a group of firefighter widows that was gaining media visibility and political power, most notably Fontana, but also civilian members of Give Your Voice. The firefighter wives felt neglected because they had not received back wages and other benefits. But more importantly, they believed the city was abandoning their husbands, who had died in the line of duty. The civilian families felt that too much attention was being paid to uniformed service personnel and city employees, and that their emotional and financial needs were not being met.[84] For both groups, though, their primary concern was that clearing the trade center site quickly would make the recovery of their loved ones' remains less likely.[85] It sickened many of them to think that their loved ones would simply be bulldozed out of the site and dumped unceremoniously on top of decades of garbage at Fresh Kills Landfill. In their view, this was morally wrong and callous. The widows lambasted DDC's Mike Burton, saying he should be made to tell the children of missing firefighters that their fathers would be found at a garbage dump. They called him "Mr. Scoop and Dump."[86]

Monica Iken, who founded the organization September's Mission to advocate for the creation of a suitable memorial to the World Trade Center attack victims, stated a few months later that she was astonished at the speed of the cleanup effort. "I don't understand why there's such a rush. It's just discouraging to see that because that's our cemetery, that's all we have. I mean, most of our loved ones are still in that site. And to see a rush [taking] place when there [are] remains that need to be found, it's disheartening because they want to just close it up like it didn't happen and call it a day. The fact that they're rushing through this process when our loved ones are still there is very hard for us."[87]

When Chief Medical Examiner Charles Hirsch described the very low probability of finding whole, or even partially intact, bodies now that the recovery effort was hitting the middle of the pile, the widows at the November 12 meeting expressed outrage.[88] Hirsch did not wish to hurt the families, or to provide cover for the cleanup effort; he was simply trying to brace them for what he thought was the most likely scenario.[89] In the following weeks, OCME officials clarified Hirsch's statement—arguing that while complete or nearly complete bodies could indeed withstand the fires that raged in the pile in the early days, smaller partial remains that typified the majority of victims would have a much lower probability of surviving and yielding usable DNA.[90] It turns out that Hirsch was technically wrong—there were a few nearly whole bodies found in pockets further down in the ruins—but at the time there was little scientific information to suggest that this would be the case.

The November 12 meeting crystalized for Giuliani, his administration, and the world at large what they already knew: the "unbuilding" of the World Trade Center was as much about emotion as it was about equipment and manpower. In order to bring about a transformation of the site and to restore New York to normalcy, they had to pay attention to the needs and desires of the families. In particular, city officials had to remember that Ground Zero wasn't just a chaotic construction site that needed to be tamed; it was also a place where nearly 3,000 people had lost their lives.

In an effort to regain the firefighters' and families' trust in the administration, Giuliani restored the search team to seventy-five and provided widows with formal channels for airing their grievances and getting information about the progress of the recovery effort. According to administration officials, revised rules would improve organization and ensure safety on the site (although very few injuries had been reported during the first two-and-a-half months of the recovery effort). In the new system, most firefighters would wait outside an active search area while a few designated "spotters" looked for signs of human remains as grappler operators and construction workers removed debris from the site. Only when potential human remains or personal effects had been uncovered could the rest of the firefighter contingent approach to investigate. The decision seemed partly due to a desire to mollify families of the missing, and partly to bring an end to the embarrassment that resulted from the protest that had taken place on November 2.[91] The tensions did not disappear, but they subsided for most of the rest of the recovery effort. Although workers in the Pit continued to bicker and even fight from time to time, interviews suggest that most low-level workers at the site had a common sense of purpose, and that most disagreements were the result of stress and frustration rather than entrenched anger or hatred.[92]

Peter Rinaldi, a Port Authority engineer who provided his professional expertise and knowledge of the World Trade Center complex to the DDC as it managed deconstruction efforts, said that most of his interactions with the FDNY were negotiations about the balance between worker safety and the recovery of remains. He noted that risks taken for one recovery might have a negative impact on many others if problems occurred, so each individual decision had to be taken in the larger context of the recovery effort.[93] Perhaps the best example of such issues was the FDNY's desire to recover remains from underneath the last major debris ramp (called "Tully Road" because it was built in Tully Construction's quadrant), which was used to get construction equipment into, and debris out of, the Pit. The material that the ramp was built on was believed to contain a vast trove of human remains because it was located in the region where the lobby of the South Tower once stood. This location was where the FDNY had had one of its command centers and where many people who narrowly missed their opportunity to escape would be found. Uniformed service personnel were anxious to excavate this area, knowing that there was at least one body there, and at one point got so close to the ramp that its structural integrity could have been compromised. DDC personnel and the FDNY had many long conversations about potential damage to Tully Road if excavations took place underneath it and the delays that this would cause to recoveries in other areas of the Pit. Ultimately, the FDNY waited until a bridge ramp was put into place to excavate the Tully Road, and numerous remains were indeed found.[94]

As the recovery effort moved forward, workers at the site and families waiting for news were beginning to accept that many of the victims of 9/11 would never be recovered—either on site or at Fresh Kills. As 2001 gave way to 2002, some blamed this absence of remains on the physical dynamics of the event (bodies had been vaporized, shattered, or burned beyond hope of identification), while others blamed it on the human limitations of the recovery effort. For some families, though, the concern was not just about whether remains would be recovered, it was about whether they would be recovered with dignity.

2

Fresh Kills

The decision to truck debris from the World Trade Center site to Fresh Kills Landfill for sifting and long-term storage would have profound ramifications for search for human remains. For the City of New York and those tasked with the unbuilding of the World Trade Center, Fresh Kills was an ideal place to bring debris from the site. For many families of the missing, moving material that potentially contained the remains of their loved ones to a landfill was an affront to the dignity of the dead. They believed that the search for remains of the missing should continue undisturbed at Ground Zero. Instead, the material would be hauled more than twenty miles away to a facility that most New Yorkers, and indeed the world, associated with decaying garbage, methane fumes, rats, and scavenging birds.

Over the next decade, controversies would develop about the way that the debris was initially searched for human remains, where and how to store the finest material once it was separated from larger pieces, and whether any human remains would come into contact with the ordinary household garbage that lay underneath the ground. Families would ultimately sue the City of New York for failing to respect their loved ones and provide them with a proper burial—taking their case all the way to the U.S. Supreme Court.

Opening in 1948, Fresh Kills had served as the city's main garbage dump for most of the second half of the twentieth century. The 2,200-acre landfill had just been closed in March 2001 and was awaiting conversion to a park when the World Trade Center was attacked on September 11. In the chaos of the rescue mission, it quickly became clear that workers would need a place to put the tons and tons of steel and debris that they were removing from the site in their frantic search for survivors and victims. There simply was no space anywhere on the island of Manhattan that could accommodate this volume of material. Despite its former life as one of the world's largest garbage dumps, Fresh Kills was seen as an ideal spot to store and sort through the debris from Ground Zero. It was isolated enough that it could be cordoned off from the public and it was big enough that all of the material could be brought to one place. Later in the effort, the proximity of the World Trade Center site to the river, and the accessibility of Fresh Kills by barge, meant that debris could be brought to the landfill by water.

Thus, in the early morning hours of September 12, 2001, material from the site—including pulverized concrete, steel rebar, asbestos, and other insulation materials—was loaded into uncovered trucks and brought to a cordoned-off 135-acre section of the landfill known as Hill 1/9.[1] There, it was unloaded and searched for evidence related to the attacks (particularly the flight recorders from the two planes that crashed into the towers), and valuables and contraband that were known to have been buried in the collapse of the towers (including guns, drugs, and money from various local and federal law enforcement agencies that had had offices at the towers), as well as remains and personal effects of the victims. The area was immediately reclassified as a crime scene. For the first month of the operation, NYPD detectives, assisted by sanitation workers, FBI agents, private contractors, and other volunteers, performed this sifting job by hand, using only rakes, shovels, and other basic equipment.[2]

As the scale and difficulty of the search became clear in late September, the U.S. Army Corps of Engineers asked the private disaster management firm Phillips and Jordan to manage the materials processing work at Fresh Kills. In mid-October, two waste recycling companies, Taylor Recycling Facility and Yannuzzi Disposal Services, were brought in to mechanize the sifting operations using giant machines normally used to separate waste for recycling.[3] In essence, the setup involved placing the debris into a series of large shaking drums that separated the material into successively smaller pieces. These pieces were then separated into debris streams of differing sizes in a way that made it easier for investigators to visually inspect all debris greater than a quarter inch. All personal effects and artifacts were examined by the FBI and sent to a "property trailer" for cleaning, indexing, and photographing. All suspected human remains were brought to the Disaster Mortuary Operational Response Team (DMORT) forensic anthropologist who was on duty at the Fresh Kills facility.[4] Because there were many restaurants and food service units in the towers, forensic anthropologists often found themselves examining nonhuman remains and food products. Anything larger than a quarter inch that was deemed to be human was individually bagged, labeled, and sent to OCME for further processing. Bits and pieces smaller than a quarter inch, which came to be known as the "fines," were collected and separated for long-term storage at Fresh Kills. What was done, or not done, with the fines, which undoubtedly contained some human remains in the form of bits of unidentifiable and desiccated bone reduced to ash by prolonged exposure to intense heat, also became the subject of controversy.

At its peak, the Fresh Kills operation employed more than 1,500 people per day and was called the "City on the Hill" because it functioned as a standalone community with dedicated phones, water service, telecommunications, heated and air-conditioned evidence examination facilities and offices, bathroom facilities, and a large cafeteria known as the "Hilltop Café," run by the American Red Cross and the Salvation Army.[5] The Fresh Kills recovery work was financed by $125 million from the federal government.

The effort, which lasted until July 15, 2002, sifted through 1.8 million tons of debris (approximately 7,000 tons per day) and found 4,257 human remains (mostly bone fragments and small, difficult-to-identify body parts, but also arms, hands, and feet—many of them manicured, suggesting that they belonged to women; 54,000 personal effects (including jewelry, credit cards, and ID badges); $76,318.47 in cash and coin; 6 kilos of narcotics; 4,000 photos; and many airplane fragments. The team also recovered 1,358 destroyed vehicles, including 102 FDNY vehicles and 61 NYPD vehicles, and 1,195 personal vehicles.[6]

All of this took place on top of fifty years of municipal garbage covered up by eighteen inches of fill dirt and giant plastic tarps. James Luongo, NYPD's incident commander at Fresh Kills, later described the inherent unpleasantness of the site. He said that the best way to cope with the stress of the mission was to learn to be able to *look* at human remains, without actually *seeing* their humanness. He described the methane that bubbled up to the surface of the site, which was produced by rotting garbage below, and the general smell of the place: "There are days that you come up here and it stinks of methane, there are days that you come up here and it just stinks of death—death has a very distinct odor. You would come up here and the entire hill would just stink, it would reek, of death."[7]

He also noted that city officials had to ask the U.S. Department of Agriculture to devise a plan to safely keep away the hundreds of seagulls and turkey vultures that circled the remains—eager to grab a bit of flesh, whether human or food. "It's not your nice peaceful view of sitting along the beach, you know, watching seagulls fly by, these are nasty birds, and they were very aggressive—we knew what piles were rich in body parts by the way the seagulls descended on it, and you would have to fight the seagulls for the human remains."[8] In the end, the Department of Agriculture settled on regular blasts of fireworks to scare away the birds. These explosions were just another layer of noise on top of the rumble of heavy construction equipment and the constant beeping of trucks backing up.

An Affront to Human Dignity

Many families were shocked by the decision to relocate the debris to Fresh Kills. Human remains being placed on a truck or barge and then dumped at a location that once handled city garbage was horrifying for these relatives. Those in the firefighting community who disapproved of the city's handling of the cleanup of Ground Zero argued that Fresh Kills was where the real "bagging and tagging" of victims was taking place. They argued that the rituals and prayers that provided respect and dignity for remains recovered at Ground Zero, whether uniformed service or civilian, did not exist at Fresh Kills.

A group calling itself the World Trade Center Living History Project argued that

> it was only at the WTC site that clergy would be called to join the recovery workers for any discovery of human remains, for important moments of respect, dignity, and prayer. Faced with inhumanity, recovery workers, most importantly, transformed barbarism into civilized behavior for the nation and the world. In these few moments there was acknowledgment of our connection to these approximately 20,000 . . . nameless, faceless parts and wholes of humans who deserved respect. No such rituals existed at the Fresh Kills Dump. No clergy, no honor guard, no flag-draped ceremony existed as a dignity at the garbage dump unearthing of friends, family or complete strangers.[9]

In the Living History Project's view, the recovery of human remains at Fresh Kills was not a symbol of an effective response to the World Trade Center tragedy or proof that the systems put in place by the city were efficiently locating these remains. Rather, it was a failure to uphold the dignity of the victims and their families. Living History Project members argued that archaeologists would never consider recovering the remains of slaves or native peoples in the way that 9/11 victims were being recovered. As such, the victims of 9/11 deserved the same respect that society accorded to other dead people—even those who died decades or even hundreds of years ago.[10] The Living History Project was only one such voice. Families were also beginning to voice serious concerns about the conditions at the landfill. Patrick Cartier, father of an apprentice electrician who was killed in the attacks, reported upon visiting Fresh Kills that it was akin to "entering hell itself."[11]

Officials associated with the Fresh Kills operation refuted such claims. While they could not guarantee that they would recover every single remain, they felt that their main objective was to recover as many as possible. They took the job seriously and did not make light of their operation in any way. NYPD Chief of Detectives William Allee told a group of reporters during a tour in mid-January 2002, "This is not a garbage dump; it's a special place. It is sacred ground to all of us. We're doing God's work and I feel honored to be here."[12]

World Trade Center Families for Proper Burial

World Trade Center Families for Proper Burial (WTCFPB) held its first meeting on December 3, 2002 and was formally chartered as a New Jersey not-for-profit corporation on November 10, 2003.[13] Kurt and Diane Horning, whose twenty-six-year-old son Matthew was killed in the attacks, founded the organization along with several other families. Matthew worked on the 95th floor of the North Tower as a database administrator for the insurance company Marsh & McLennan. He worked on a team that ensured that the company's data could be quickly recovered in the event of a disaster or catastrophe. Matthew's wallet and a piece of his skull bone had been recovered through the Fresh Kills operation and the Hornings believed that more of his remains were "buried amidst the debris at Section 1/9 at Fresh Kills."[14] Horning was on a tour of the Fresh Kills facility led by FBI Special Agent Richard Marx in July 2002 when it hit her that the fines, the bits of material less than a quarter inch, "contained Matthew's and others' human remains." Indeed, according to her affidavit, it was Special Agent Marx who informed her that the fines contained at least some cremated remains.[15]

The primary goal of WTCFPB was to ensure that the fines were separated from the rest of the World Trade Center debris and were not permanently intermixed with garbage at the Fresh Kills landfill site. The group wanted this material to be placed in special containers and ideally returned to the World Trade Center site, where they would be permanently interred at the place of death. If this outcome was not possible, then they hoped that the fines would be placed in a suitable "cemetery and memorial garden which serves no other purpose than to bury our dead with respect and to provide the families and friends with a place for quiet reflection."[16] Fresh Kills Landfill, in their view, simply did not fit the bill. It was, inverting Lincoln's description of the cemetery in Gettysburg, "unconsecrated ground."[17]

WTCFPB emerged from a support group for families who had lost loved ones in the September 11 attacks. None of the core members had been politically active before September 11. They came from ordinary, middle-class families from New York and New Jersey and were not prepared for the shock and tragedy that befell them that day. Early on in their advocacy work, the Hornings and their partners spoke primarily in terms of dignity and respect for the dead, but as they became more savvy activists, and enlisted the helped of noted civil libertarian lawyer Norman Siegel, their language shifted to the more legalistic concepts of property and ownership. The core of the group consisted of eleven families, but on December 20, 2003, WTCFPB created an online petition that asked members of the public to support the organization's cause and to note whether one of their loved ones had perished in the 9/11 attacks. Based on the results of this survey, which approximately 62,500 people signed, WTCFPB claimed that it represented around a thousand affected families.[18]

In addition to the fate of the fines, WTCFPB members were also concerned that, in the first month of the operation, debris that potentially contained human remains was "bulldozed" over the edge of the hill and mixed in with household garbage in a location in Section 1/9 known initially as "the North Field" and then "Area A." The Department of Sanitation admitted that they did this in order to make room for newer loads of debris.[19] WTCFPB contended that this debris was "suddenly subjected to foraging by droves of seagulls attracted by the human parts contained in it."[20] Officials in charge of the Fresh Kills investigation insisted the material was had been searched in the first few weeks of the operation, and that it was excavated and resifted at the end of the operation. WTCFPB contended that this was untrue.[21] Analysis of bone material recovered from the rooftops of buildings around Fresh Kills (i.e., bones removed by seagulls from the site or from barges and trucks arriving at the landfill) demonstrated consistently that none of them were human in origin.[22]

In their complaint, WTCFPB acknowledge that Taylor and Yannuzzi did mechanically sift some of this material and "recovered numerous body parts and human remains," but that they were neither authorized nor instructed by agents of the city to do so.[23] They claimed that approximately 414,000 tons of debris went unsifted, leaving "at a minimum hundreds of human body parts of victims of the World Trade Center" buried at Fresh Kills. In a later affidavit, James Taylor, the retired chairman of Taylor Recycling Facility, provided his own calculation, claiming that approximately 223,000 tons of debris was never sifted at Fresh Kills.[24] According to his testimony and the affidavit, Taylor believed that the material bulldozed off the side of the North Field/Area A part of Section 1/9 was never fully resifted and therefore contained human remains that had not been recovered. "This is why these family members are so unsettled and why Diane Horning doesn't sleep at night," he said. "It breaks my heart to know that Diane is right about the issue of September 11th human remains."[25]

WTCFPB negotiated with the city over the material at Fresh Kills for many years. Horning and other family members regularly asked FBI and City of New York officials about the status and fate of the fines. In all of their conversations, officials at Fresh Kills reassured families that the fines were being kept separate from household waste and the rest of the debris from Ground Zero. WTCFPB claimed that Fresh Kills director Dennis Diggins assured them in August 2002 that ground asphalt millings—recycled road material—had been put down where the fines were being stored in order to provide an additional barrier between the material and the waste below the topsoil. While onsite during a July 2002 visit, FBI Special Agent Richard Marx told them that they "could request the fines [for proper burial] when the recovery effort was concluded."[26]

WTCFPB families wrote to the Office of the Mayor and the Department of Sanitation after their conversation with Marx to find out what the city planned to do with the fines, and to ask that they be treated with respect and properly buried. Mayor Bloomberg's liaison to the families of 9/11, Christy Ferer—whose husband Neil Lavin, the executive director of the Port Authority, was killed on 9/11—responded quickly to the families, assuring them that the material was being maintained with reverence and that one plan would be to use the debris as fill material in the redevelopment of the World Trade Center site, but that it would be costly. She noted that she was "reaching out to private money to see if we can finance its move if and when that becomes an option." It is interesting that although the families were primarily interested in the fines, estimated to be approximately 360,000 to 480,000 tons, Ferer refers to all the debris sent to Fresh Kills, including material that most likely contained no human remains. In a clear effort to empathize with the families, she wrote, "The closing of Fresh Kills [i.e, the conclusion of the sifting operation there] has rekindled raw emotions in all of us. It is so painful to imagine the unidentified remains of our loved ones intermingled with the debris that remains at the landfill. I am sensitizing City Hall to the fact that this debris cannot be forgotten and they have to plan for its future."[27]

The families, however, had reason to distrust Mayor Bloomberg. On numerous occasions, both private and public, Bloomberg expressed a lack of personal interest in funereal rights and prevailing notions of proper burial. Beginning in 2002, WTCFPB requested a meeting with Mayor Bloomberg to ask him to support their efforts to remove the fines from Fresh Kills and provide what families considered to be a more fitting resting place. According to Diane Horning and news reports, however, when Bloomberg finally met with the Hornings in April 2003, he was dismissive of their request and seemed puzzled by their continued passion about the human remains at Fresh Kills. He told the families that he had little interest in the corporal remains of humans after death. He told them he intended to leave his own body to science and that he had visited his father's grave only once. When Kurt Horning asked him if he would be bothered if his own father's remains were kept at Fresh Kills, he said he would not. In the end, Bloomberg refused to endorse their mission, and when the Hornings pointed out that many other politicians, including senators Schumer and Clinton, had done so, he responded to the effect that it's very easy to get important people to write letters on your behalf, but much harder to get them to do something for you. In other words, Bloomberg felt that he was just being honest when he said he would not actually do anything to help them.[28] Bloomberg cut their meeting short in order to deliver a civic award at the halftime of a New York Knicks basketball game at Madison Square Garden.[29]

Bloomberg, it turns out, was only partially right in his assessment of the intentions of politicians. WTCFPB families succeeded in working with members of the New Jersey legislature to pass a bill in December 2003 ordering the Port Authority to move the fines out of Fresh Kills, which was signed into law by New Jersey Governor James McGreevey. But because New York State never passed a similar law, none of the New Jersey law's provisions were ever enforced. Despite WTCFPB's efforts to sway Governor Pataki and other New York State officials, a New York version of the bill never made it through the state senate.[30]

Ferer, while being sensitive to the needs of families, felt strongly that they had to be balanced against the needs of businesses and residents of lower Manhattan, who were eager to get on with their work and lives. Hence many of the World Trade Center families came to believe that her role was to present the mayor's needs to the families rather than vice versa. In an editorial in the *New York Times* in 2002, Ferer made a point of contrasting the "silent majority of families" who send her messages that are "moderate" in tone, mostly about wanting to get on with their lives and never return to Ground Zero, with those presumably more extreme families who were obstructionist and belonged to organized groups that made demands to the city about what should and should not happen at Ground Zero.[31]

The silence of families of 9/11 victims, however, did not mean that they approved what was taking place at Fresh Kills. For many relatives I spoke with, the idea that a loved one's remains were in the landfill was simply too horrendous to think about. Monika Iken, who has been perhaps the most important family advocate in the creation of the memorial at Ground Zero, told me in 2011 that it was an issue that she did not have the stomach to deal with. While she was deeply appreciate of the work of WTCFPB, and supported it in any way she could, she avoided thinking about the possibility that her husband Michael's remains were there:

> Thank God somebody else is dealing with that. . . . The whole idea that he is in Staten Island and not over here [in lower Manhattan] and he is in a dump—the whole thing throws me over the edge. I cannot even conceptually put myself there. . . . I have never been there, I have never seen it, nor do I want to. I cannot handle it. So what I do is avoid it so that I can't—my Michael is not there. That is how I look at it.[32]

For many families whose loved ones were killed on 9/11, the pronouncements from Bloomberg and his staff suggested an insensitivity that was both inappropriate and emotionally damaging. Many family members who were not passionate about the human remains issue—because they did not believe that their loved were in the fines at Fresh Kills, because it did not matter to them, or because they simply could not handle thinking about the problem—felt that the city was treating WTCFPB families unfairly. Nikki Stern, whose husband was killed on 9/11 and who was the executive director of Families of September 11, argued that "more compassion and more respect could have been shown to these families early on. The issue is not whether I personally believe that part of my husband is at Fresh Kills. . . . The issue is, there are family members who do believe that. How are they being treated, and how were they treated? People who have this belief are not crazy and should not be treated as such."[33]

In 2004, Horning and other family members at the core of WTCFPB began to attend New York City Planning Commission hearings in order to monitor the city's plans for creating a park at Fresh Kills, to voice their opposition to keeping the fines at the landfill, and to demand a new search of the material they argued had never been properly sifted. WTCFPB also sought to make it easier for families to visit Fresh Kills to pay respects to their deceased loved ones.

It is unlikely that many families would have wanted to visit the Fresh Kills site, but WTCFPB felt that they had a right to do so. WTCFPB believed that it was far too difficult to get to the site. There was no public transportation to Fresh Kills, visitors had to request official permission, and an escort from the Department of Sanitation was required. Visitors were also required to sign a waiver excusing the city from liability should any injuries occur during the visit. For Horning, though, it was the starkness of the place that made the visits so difficult: "The current mourning site is not a place to bring children, the infirm or the elderly. The walk itself is difficult as the soil erosion can be as deep as four feet. There is no feeling of solace or closeness to your loved one. It is a formidable place. Some days the smell of methane is too much to bear. During visits, some of us have found old tires, carpeting, construction fill, pieces of blistered steel, glass, bolts, and fire hose, shoes, and tiling at the site. It is simply devastating."[34]

WTCFPB families felt that Bloomberg and his administration saw them as a nuisance and an impediment to progress in the city's recovery rather than an interest group with a legitimate concern.[35] For the men and women of the law enforcement community who had devoted ten months of their lives searching for human remains at Fresh Kills, such claims and allegations were frustrating and offensive.

In interviews and in his affidavit, NYPD Fresh Kills Incident Commander Luongo noted that he personally witnessed the collapse of the towers when responding to the attacks and understood the significance of the operation for the families. He stated that if any of them had any reason whatsoever to question the quality of the search for human remains, he would have "halted the operation immediately."[36]

Luongo further said that no workers at Fresh Kills had ever complained to superiors about improper searching and that he believed that they would have done so given the number of their colleagues who had perished on 9/11.[37] He also noted that search personnel were briefed on proper procedure before the start of every shift, and that forensic experts regularly reminded workers about what to do in the event that suspected human remains were found.

Luongo explained that he personally witnessed the excavation and sifting with the Taylor and Yannuzzi machines of any material searched by hand before the equipment was available. He explained that Department of Sanitation workers used GPS to make sure that they were excavating the correct material and searched below the officially recorded bottom of the debris field in order to ensure that they did not miss anything relevant. Finally, he pointed out that any material that was not subjected to the Taylor and Yannuzzi machines was simply too large to fit in it. This debris was searched carefully by hand. Therefore, there was no WTC debris that could have been screened but was not. These claims were confirmed in great detail by Dennis Diggins, director of Fresh Kills for most of the search effort and who supervised the excavation and sifting of the early debris in May and June 2002, and by Scott Orr, who led Phillips and Jordan's operations at Fresh Kills and was stationed there full time from October to August 2002.[38]

WTCFPB sought permission to take drill bore samples at their own expense, but the city did not allow them to do so. In WTCFPB's view, these samples would have allowed them to definitively prove their case, but the city believed such testing was unnecessary given the quality control measures already in place.[39] Thus, it is impossible to know for certain which side is correct.

Beginning in early 2002, the Hornings and other families concerned about the fines at Fresh Kills heard rumors that the fine material less than a quarter inch was only separated temporarily and that the Department of Sanitation "apparently recombined the fines with debris from the World Trade Center and dumped them into the landfill."[40] They also claimed they had discovered that no millings had ever been laid to separate WTC debris from fifty years of municipal waste. Additionally, they took issue with the argument that it would be expensive to move the fines back to the World Trade Center site, noting that the city estimated that it would cost more to move the fines than the entire sifting effort at Fresh Kills ultimately cost. According to the concerned families, the city had "exaggerated the difficulties and cost of relocating the fines in an effort to thwart [their] efforts to ensure a proper and decent burial for their relatives."[41] Later, the families were concerned that the emphasis on speed above all else may have infected the operation at Fresh Kills the way it had created incentives to clean up Ground Zero as quickly as possible.

Perhaps most fundamentally, there was a disagreement between the city and WTCFPB about the extent to which the fines ought to be considered human remains. For the city, every reasonable effort had been taken to retrieve any bit of human tissue or bone larger than a quarter inch. While it was undeniable that some fraction of the fines was human in origin because of the crushing nature of the collapse of the towers, this did not mean that all of the fines, which was mostly pulverized concrete and other building material, needed to be treated the same way one would treat a body or a body part. At some point, city officials believed, one had to decide that enough was enough and accept that some small quantity of human remains would end up in Fresh Kills.

For WTCFPB, the existence of human remains in the fines represented not just some abstract presence, but rather the potential presence of *their* loved ones. The fines themselves became corporeal.[42] As James Taylor, the chairman of Taylor Recycling Facility who became a vocal advocate for WTCFPB, stated in an interview with the *New York Times*, "You bring tears to my heart when you make me talk about this [the fines], but [are] there human beings in that powder material? Absolutely. There's 2,749 spirits theoretically in that fines material."[43] As such, it was imperative that this material be treated reverentially and accorded the same respect that one would give to human bodies or body parts. Given the comingled nature of the human remains that were present in the fines, WTCFPB wanted the entire volume of fine material to be removed from the landfill to prevent the permanent internment of the possible remains of their, or another family's, loved ones in such a disrespectful environment.[44] The fact that the landfill would one day be turned into a park did not comfort these families. Fresh Kills' past life as a garbage dump could not be erased by landscaping and environmental remediation.

In an e-mail message to Diane Horning, Diana Stewart, whose husband Michael was killed in the collapse of the North Tower, linked the reverential treatment of remains at Fresh Kills to an act of defiance against terrorism itself:

> For those like us who received partial remains, and the many who have received none, the remains at Fresh Kills require a proper, American burial, with all the dignity, ceremony and prayer which would accompany a burial if we could identity ash from ash and put a name on each. These unidentifiable remains are literally our kin, loved ones, fellow Americans. And for those who were none of these, we still embrace them equally, as our country always has, giving equal opportunity to those who come here to join our way of life, in peace.
>
> Please give all these unclaimed remains the dignity of burial, and I will be among the mourners who will stand to bear witness, express eternal gratitude to those who worked to free the remains from the rubble and, most importantly, stand in defiance of those cowardly murderers who thought these lives unworthy . . . thereby proving that in our country, every life matters.[45]

Thus, for Stewart, any unidentified human remains were the collective property of each and every family that lost a loved one in the WTC attacks, and that it was the duty of these families and the nation to treat each and every bit of human remains as if it belonged to a known victim of the tragedy.

On September 29, 2003, Mayor Bloomberg unveiled long-term plans to turn the landfill into an expansive park and nature preserve with both recreational and natural spaces, as well as a memorial of some sort to the victims of the World Trade Center attacks.[46] Implicitly acknowledging that there were human remains at the site, he suggested that a few handfuls of the fines would be brought to Ground Zero and incorporated into the reconstruction of the site. As plans for the site developed, it became clear that the Bloomberg administration envisioned two memorials at Fresh Kills—one involving the towers, which would be represented by two long side-by-side earthworks the length of the towers themselves, and another for the victims of the attacks, represented by a mound containing the fines. While administration officials viewed this as a dignified way to respect the dead, the families of WTCFPB disagreed. They simply did not want any human remains entombed atop five decades of New York City garbage.

On March 24, 2004, the City Planning Commission held a meeting about the plan to turn Fresh Kills into a park. Several WTCFPB families attended that meeting and stated that they were against any attempt to turn the human remains into a permanent feature of this park.[47] On July 19, 2004, Jonathan Greenspun, the head of the mayor's Community Assistance Unit, informed WTCFBP that the fines would not be moved from Fresh Kills, even though he knew this decision would disappoint the families. Greenspun did, however, welcome them to participate in the planning of a memorial at the park. Although the city claimed that no alternative sites were available to store the fines, the families contended that they knew of at least two appropriate sites: Liberty State Park in New Jersey and a non-landfill site in the Fresh Kills area at Muldoon Avenue.[48] Later they would suggest Governor's Island as a third alternative.[49]

In addition to the general concerns over leaving the fines at Fresh Kills, WTCFPB families were concerned that their regular visits to Sections 1 and 9 of Fresh Kills landfill demonstrated that these areas was not being treated with reverence—rather they were littered with "chunks of steel, shoes, tires, carpets, portions of storm drains and other garbage. In addition, the area bears signs of erosion and tracks from large construction vehicles."[50] For members of the Bloomberg administration and others not involved with WTCFPB activism, such claims were "misconceptions" about what the landfill looked like more than two years after the recovery mission ended. They noted that only "clean" construction fill was used to cover the rubble from the WTC.[51]

From August to October 2004, WTCFPB attended several City Planning Commission meetings, requesting that all potential human remains be recovered before the landfill was turned into a park. They also notified the commission that those in charge of the recovery effort were not cleaning up the site, but merely covering it with an additional layer of dirt to conceal the comingling of the fines with garbage and other debris.[52]

Beginning with their October 2004 Notice of Claim against the city, WTCFPB changed their tactics from negotiation about dignity and respect for the dead to assertion of familial and property rights over the unidentified remains of their relatives at Fresh Kills. In a March 11, 2005 letter to Mayor Bloomberg summarizing the families' position, attorney Normal Siegel wrote:

> Federal courts have recognized that next of kin have property rights in the remains of deceased relatives that are entitled to due process protection. . . . The law makes no distinction between cremated remains and an intact body. . . . What is of immediate importance is a determination of whether the City, and its representatives, has afforded [victims' families] an appropriate level of due process protection in deciding that Fresh Kills is the only appropriate place where the remains can be permanently kept. Based on our research, we conclude that the City has systematically ignored the civil and constitutional rights of our client.[53]

Siegel then went on to accuse the city of not dealing with concerned families "in good-faith" because they were not always invited to meetings about Fresh Kills or given accurate information about what was going on there, and, further, that they lied to families about the practicality of moving remains. If the city did not address the families' concerns, grounded in the legal right to control the disposition of the remains of direct kin, the families would be forced to pursue legal action against it.[54]

While some media outlets covered the possible legal action sympathetically, the *New York Post* used it as evidence that the families of WTCFPB were no longer engaged in a rational debate over the remains.[55] An editorial in the newspaper noted that "the city's proposal is sensitive and appropriate—not perfect, but hardly reprehensible. But the families are in no mood to compromise—they consider their demands non-negotiable. Following through with a lawsuit may prove emotionally satisfying for the families, but it's unfair. The city has behaved both honourably and respectfully. It's long past time to move on."[56] In another editorial, Andrea Peyser noted that "these relatives just can't let go. And perhaps they never will. The latest fight to be waged by 9/11 families . . . threatens, I fear, to weaken their credibility in the eyes of a public that is, frankly, growing weary of them and their many complaints."[57]

When intense discussions failed to produce an outcome that satisfied Horning and other WTCFPB families, they filed suit against the City of New York in the United States District Court for the Southern District of New York in August 2005. This court was responsible for hearing all litigation related to the World Trade Center attacks as a result of a provision in the federal Air Transportation Safety and System Stabilization Act.[58] Their legal strategy rested on the claim that families of the victims had a property right in the remains of their loved ones under both federal and state law. Specifically, they argued that next of kin have the right to "take possession of the remains, to make arrangements for a proper and decent burial, and to give instructions regarding the disposal of remains." These rights, in their view, extended from intact bodies all the way to "body parts, bone fragments, small tissue particles, and cremated remains."[59] They further claimed that their due process rights had been violated because city authorities (1) misled them about the treatment and storage of human remains at Fresh Kills, particularly in the first month of the operation; and (2) by "bulldozing topsoil over the area containing remains" the city "set in motion a process that will make the remains a permanent fixture of Fresh Kills."[60] Finally, they noted that New York State law prohibited the interment of human remains at Fresh Kills landfill.[61]

The case would ultimately be heard by Judge Alvin K. Hellerstein, who presided over nearly all litigation emerging from the September 11th attacks, and who went to great lengths to empathize with those who had been killed or injured in some way in the attacks or recovery effort.[62] In the months after WTCFPB filed suit, however, Hellerstein ordered the families and the city to try to come to some agreement in order to avoid a lengthy and emotionally damaging trial. He even arranged for a series of less formal conversations to take place in his chambers in an effort to avoid turning the issue into a purely legal dispute. He recommended that the families think more symbolically about the fines, since it was unlikely that the entire amount would be transported from Fresh Kills. Perhaps, he suggested, a portion of the fines could be brought to another site as a symbolic memorial to those who had died on 9/11.

For the families, however, the notion that the fines be treated symbolically was akin to accepting the idea that their loved ones were also only symbolically dead. Their loss was total, and they were not going to compromise on what they perceived as the right way to handle their loved ones. Indeed, after Hellerstein described the positive attributes of a park at Fresh

Kills in open court, Laura Walker, the wife of a victim of the attacks, stood up in the gallery and chided the judge for even suggesting that the remains be kept at the landfill. "You should be ashamed of yourself," she told him. In keeping with his efforts to find some way of resolving the dispute without formal litigation, Hellerstein responded to her outburst not by asking her to be removed from the courtroom, but by apologizing to her and other families for their pain.[63] Despite Hellerstein's demeanor and desire to reach a compromise, one could not be found at this meeting or in subsequent discussions.

Disappointed but undeterred, WTCFPB moved forward with collecting a wide-ranging mix of evidence for a full-scale legal case that would prove negligence at, and mismanagement of, the Fresh Kills facility. In an affidavit submitted as part of the WTCFPB suit, for example, a unionized Taylor employee at the site named Eric Beck claimed that P&J, the private contractor, constantly told workers to "keep the tonnage up" (i.e., the amount of material sifted per day) and to make the belts run as quickly as possible.[64] At the same time, he said that NYPD officers examining the debris as it went down the conveyor belts were constantly asking workers to slow the process down so that they could adequately search the material that passed them.[65]

In another affidavit, retired NYPD officer John Barrett, who spent a few shifts at Fresh Kills during the first six weeks of the operation (before mechanical sifting began), claimed that Sanitation Department workers took material away from him before he had had the chance to fully examine it and that NYPD officials at the site failed to properly investigate his allegations.[66] Barrett also claimed that the steel beams removed from debris for recycling were not always checked for human remains and other artifacts. In cases where they were actually searched, the location of any relevant material recovered from them was not clearly recorded.[67] This made it difficult to determine whether or not all, some, or very little of this material was actually searched.

Theodore Feasor, the director of mechanical operations at Fresh Kills, was responsible for ensuring that the recovery effort had the necessary equipment. He was concerned that material recovered from the debris was not being rinsed before inspection in order to ensure that human remains and other artifacts weren't being missed because they were difficult to see. He stated that his suggestion to do so was rejected by his supervisors at an October 2001 meeting because it would cost more and slow down the process.[68] He also doubted that the material deposited in North Field/Area A was ever resifted at the end of the search effort. As such, he was "absolutely convinced" that "hundreds of human body parts and human remains" had gone undiscovered.[69]

Most shockingly, Eric Beck claimed that he observed New York Department of Sanitation workers "taking . . . fines from the conveyor belts of our machines, loading it onto tractors, and using it to pave roads and fill potholes, dips, and ruts" at the site.[70] Such a claim must have been upsetting to families concerned that these fines contained fragments of their loved ones—a possibility that was given official backing in a 2003 letter from Charles Hirsch to Diane Horning stating that he was "virtually certain that at least some human tissue is mixed in with the dirt at the Staten Island landfill."[71]

Beck's credibility, however, was at least partially undermined by his claim that the operation recovered approximately 2,000 bones per day during the first few months of the job. Given that only 4,000 bones total were recovered from Fresh Kills (compared to approximately 16,000 at Ground Zero), his claim either represents a faulty memory or a misunderstanding of what was happening at Fresh Kills during the year he was there. One reasonable explanation is that he mistook animal products for human remains.

In a supplemental affidavit for the defense, Dennis Diggins vehemently denied this claim and others made by Beck, Feasor, and Barrett. He claimed that all three overstated their responsibilities at Fresh Kills and made allegations based on incomplete knowledge of the search process—in other words, that none of these witnesses had any real credibility. In particular he said the men failed to realize that not all of the material that was not sifted in the Taylor and Yannuzzi machines was steel. There were other things that simply could not be broken down into small enough pieces, and these chunks were carefully searched by hand.[72] With respect to Beck's claim that the fines were used to pave roads and fill potholes at Fresh Kills, Diggins argued that the "accusation is such a perversion of the truth, and is so far-fetched, that it should be discounted completely without further comment. However, because it is so offensive to me personally, and to others at DSNY, I feel I must respond to it." Diggins went on to note that there were two sources of millings and crushed stone for road projects at Fresh Kills—both of which provided more than enough material to construct and maintain any and all roads necessary for the search activities at Fresh

Kills.[73] With respect to Feasor's complaint about the failure to wash down debris with water, Diggins noted that the idea was rejected specifically because OCME stated that such wetting down would further compromise already damaged DNA and make it even more difficult to identify the source of the remains. Cost and speed were simply not issues in the decision.

In an effort to dispel notions that the North Field/Field A was not resifted toward the end of the recovery effort, Diggins supplied minutes from several meetings that made explicit mention of the excavation and sifting of this area. There were also mentions of resifting material underneath the sifting equipment to ensure that nothing was left behind in these areas once the operation was complete.[74]

No matter how much evidence was brought to bear on the recovery operations at Fresh Kills, however, the WTCFPB families were unconvinced. In their "Statement of Undisputed Material Facts in Opposition to Defendant's Motion for Summary Judgment," for instance, they met each and every claim made by the city and their contractors with the statement "Plaintiffs lack information sufficient to form a belief as to the truth of these statements." They continued to reassert the veracity of the statements made by workers who had submitted affidavits on their behalf. They neither directly responded to the counterarguments made by the defendants nor mentioned any of the evidence provided by the defendants to rebuff the claims made by Beck, Feasor, and Barrett (most obviously the meeting minutes making it clear that at least some excavations of the North Field/Area A were taking place throughout June 2002).[75] They did request additional operational meeting minutes besides those provided by Diggins, but said they never received them.[76]

The WTCFPPB families were seeking something greater than facts about the recovery process—they were looking for justice for their loved ones and truth from a city government that seemed to want them to go away. Their experience of being mothers, fathers, brothers, sisters, and spouses to the dead was the issue that needed to be addressed—the concerns they had with the Fresh Kills sifting operation were secondary to, and in service of, this concern. Indeed, in explaining the "heart of the issue" that defined WTCFPB's existence, the organization quoted from an "unnamed child" whose father had died in the World Trade Center attacks: "My Daddy is not garbage."[77] They wanted to be treated with respect and they wanted *their* wishes to be honored. For Bloomberg and the city, though, they were but one constituency—albeit a constituency that had become skilled at using the media to turn the Fresh Kills issue into a "political football."[78]

Oral arguments in the case took place on February 22, 2008. Judge Hellerstein began by noting that he had tried hard to broker a negotiated agreement between the two parties—but that as a judge he simply could not meet both sides' demands. Throughout the proceedings, and in his final ruling, he made it clear that the purpose of the law is to help society function the best that it can, not to enact perfect justice. He opened the hearing by stating, "To those who lost their loved ones in 9/11, in a very deep way there can never be justice. There can never be the return of a son or a daughter or a husband or a wife or a lover or a child."[79]

On the other hand, Hellerstein noted, the people responsible for the cleanup had two jobs: to remember and respect the dead, but also to repair the hole in the city and thus enable it to continue to function. Hellerstein stated that when a compromise could not be reached, "we come back to what the court of law can do—to read a complaint, to evaluate an answer, to consider motions and briefs and legal discussions and come back to issue a ruling on which the Court is sure will satisfy neither the plaintiff nor the defendant."[80]

The hearing first addressed the question of WTCFPB's standing to bring the case and whether they actually represented the thousand-plus families that signed the petition. Hellerstein wanted to grant the organization standing to raise the question of whether they had a property right in the remains at Fresh Kills as next of kin, but in order to do so, they would have to show that the city had violated their constitutional rights. The city's lawyer, James E. Tyrrell, Jr. vehemently challenged the standing of the group. In addition to the issue of whether WTCFPB could be definitively linked to particular remains at Fresh Kills, Tyrrell also questioned the extent to which the group actually represented the true voice of families of victims—in that the vast majority were not actually members of the organization. The thousand-plus families represented had simply signed an online petition. Tyrrell wondered whether there was an equally large number of families who "do not want the hallowed ground on which the [fines] are located [i.e., the Fresh Kills landfill] to be disturbed," but who had not yet been surveyed.[81] He also questioned whether the court had the authority to order the City of New York to redo its sifting operation and move the fines to a new location as dictated by a small group of families. Tyrrell further noted that, although the claims being made in this case were individual, the relief requested was collective.[82]

For Tyrrell, the fact that WTCFPB did not have a large body of active members was damning, but for Siegel, the lawyer for the claimants, the organization's size was merely a byproduct of its inability to poll all 2,749 WTC families because their contact information was tightly controlled by the city. Besides, Siegel noted, New York is a "town of big mouths" and families who did not belong to WTCFPB would speak up if they disagreed with the lawsuit.[83]

Hellerstein set out three tests that needed to be passed in order for him to rule in favor of WTCFPB: "The first is the existence of a constitutionally protected property or liberty interest; second, the deprivation of the interest by the defendants; and third, that the deprivation was without due process of law."[84] He then pointed out that WTCFPB was claiming a property interest in the remains, but that there were no particularized remains in this case. "I mean, that's a tragedy," Hellerstein lamented, "but it is also the constitutional hurdle."[85]

Siegel countered that his clients lacked a definitive property claim because of decisions made by the defendants over which the plaintiffs had no control. Hellerstein stated that there had to be actual remains, not a theoretical or speculative claim that such remains exist. Siegel repeated that the lack of identifiable property was a direct result of decisions made by the state that "shock the conscience" and therefore require redress.[86]

Hellerstein did not find this line of argument plausible. He noted that the city felt it had no choice but to move the debris from the World Trade Center site to Fresh Kills. As such, it could not be found guilty of violating the constitutional rights of families of victims of the attacks. For Siegel, the imperfect sorting in the first thirty-two days, combined with the claim that fines were being segregated while they actually were not, amounted to a constitutional violation. For Hellerstein, though, the city was performing a governmental function in clearing the streets of debris and sorting it out at a waste disposal facility. However inadequate the city's conduct may have been, it did not rise to the level of a constitutional violation.[87]

Whether WTCFPB had standing or not, Hellerstein seemed eager to hear their case in full, partly as an act of respect for the families and their loved ones and perhaps in an effort to end the controversy, which was becoming increasingly public. The question that needed to be answered was whether the fines were human remains, as the plaintiffs alleged, or merely the "undifferentiated material that slipped through a quarter-inch sieve," as Hellerstein rather crudely put it.[88] In Hellerstein's view, the existence of human remains in the fines, however likely it may be, was still speculation. He said, "the City regarded it as debris and the City regarded it as something that had to be dealt with and the City may well have been wrong and may well have been callous and may well have been indifferent but did they commit a constitutional violation of due process law?"[89]

In his argument, Tyrrell reminded Hellerstein that in order for anyone to have a property right in remains, they must identifiably belong to the person seeking control. In this case, the plaintiffs were not even asking for forensic identification of the remains. This suggested, according to Hellerstein, that they "feel there are remains in the [fines] but the remains are not identifiable."[90] Tyrrell and Hellerstein both agreed that this made the case unique—that people were asking for control over human remains that might, but did not definitively, belong to their loved ones. Further, they were doing so on behalf of collective interests rather than their own. Tyrrell also added that WTCFPB had to prove deliberate bad intent on the part of the state in order to demonstrate a constitutional violation in this case.

As the oral arguments came to a close, Hellerstein signaled that it would be very difficult for him to rule in favor of the plaintiffs (he noted that the "chances to dress up this argument in constitutional clothing does not look too bright"), and that he knew his ruling would not bring solace to the plaintiffs. In essence, he asked them to go back to the negotiating table, because he was not able to give them what they wanted. "I would like to see more effort renewed in coming together on some kind of settlement," he told both sides. "I would like to think," he concluded, "that if we can resolve this issue amicably we would better achieve everybody's purposes than to have another opinion come down in a fat law book and be appealed and appealed again and forever grieved on."[91]

Hellerstein began his written opinion in the case with the following description:

> The terrorists of September 11, 2001 murdered 2,749 people in Towers One and Two of the World Trade Center. Approximately 1,100 of the victims perished without leaving a trace, utterly consumed into incorporeality by the intense, raging fires, or pulverized into dust by the massive tons of collapsing concrete and steel. Full bodies were recovered for only 292 victims, and partial remains were found for

> another 1,357 victims—sometimes a fragment of bone or a possession, sometimes more. City workers and contractors have inspected every bit of debris and, using sophisticated equipment, sifted the particles of debris to the extent of one-quarter inch of diameter, the space between the concentric circles of a small paper clip, with no further success. All human remains that could be identified, were identified. Only dust remains.[92]

In the end, Hellerstein agreed with the city that there was no clear property right in this case because any remains at Fresh Kills were unidentifiable. He did not accept the plaintiff's claims that the recovery effort at Fresh Kills left hundreds or thousands of identifiable human remains in the landfill. While the city may not have been as careful as it could have been, its actions fell far short of violations of human conscience.[93] Further, any violation of the plaintiffs' religious sensibilities by the city government was unintentional and a byproduct of its desire to get lower Manhattan back on its feet. In what can only be described as a lack of sensitivity to how WTCFPB families would read and hear the language in his opinion, Hellerstein imported his statement from oral arguments that the fines were "an undifferentiated mass of dirt" to which no property rights could be attached.[94] This description of what WTCFPB believed to be the remains of loved ones continued to anger those families who were most concerned about the status of the fines at Fresh Kills.

This continued pain did not surprise Hellerstein. In the conclusion of his opinion, he urged the families of WTCFPB to accept the fate of the fines and to stop fighting for their removal. He asked them to channel their anger and frustration not against the city, but toward working with government officials to build a memorial at the landfill site to honor their loved ones. "The City has a plan for a beautiful nature preserve and park at the Fresh Kills site. There is room for a memorial on a height with a view of where the twin towers stood. The energy applied to this lawsuit might well be transferred to participating in the planning of the park and memorial. What better reverence could there be than a memorial that both recalls those who died, even without leaving a trace, and points to the tenacity and beauty of life that must go on? The terrorists sought to destroy our lives and our freedom. They failed, and a memorial in such a beautiful setting can symbolize the vital continuation of our vibrant democracy."[95]

Obviously unhappy with the decision, and unwilling to accept that their loved ones' remains could be classified as an "undifferentiated mass of dirt," WTCFPB appealed Hellerstein's ruling, but to no avail. The United States Court of Appeals for the Second Circuit found that Hellerstein committed no errors in rendering his decision and that the city had not violated any constitutional rights of the plaintiffs. It noted that the city was forced to deal with a situation that it had never faced before, that it did the best that it could under the circumstances, that its agents acted with the best of intentions, and that any mistakes made were the result of uncertainty and the desire to quickly return the city to normal rather than a malevolent or reckless intent of the government. Further, any shock to the conscience was caused by the magnitude of the event rather than the response of local officials.[96] The appellate court concluded with an echo of Hellerstein:

> On a human level, plaintiffs' claims are among the most compelling we have ever been called on to consider. They have endured unimaginable anguish, and they seek nothing more than the knowledge that their loved ones lie in rest at a place of their choosing. We regret that we cannot bring them solace but we echo the sentiments of the District Court: "The events of September 11, 2001 will never be forgotten. No one knows the truth of these words more than those individuals who lost their loved ones to the attacks. In a very real sense, those individuals have suffered a wrong for which there can be no remedy. No matter the authority or power of this Court, it cannot bring back the loved ones lost, and it cannot bring peace to the plaintiffs or surcease to society's collective grief around the events of September 11, 2001."[97]

On June 1, 2010, WTCFPB petitioned the U.S. Supreme Court to hear an appeal of this latest ruling, but the request was rejected on October 4, 2010, ending the group's efforts to have the fines removed from Fresh Kills and the North Field/Area A debris resifted.[98]

Although the plaintiffs were ultimately unsuccessful in gaining a say in the fate of the fines from Fresh Kills, they continued to fight for what they perceived to be the dignified handling of remains from the World Trade Center tragedy, both at the former landfill and at the site. During the Fresh Kills fight, they attracted the attention of the media, appearing regularly in local, national, and international newspapers, magazines and television news reports. They would ultimately join forces with other World Trade Center activists and use their collective media connections to stir up the two controversies that will be discussed in the second part of this book: first, over the meaning of the discovery of additional human remains around lower Manhattan several years after the recovery operation at the WTC had been declared over; and second, how unidentified and unclaimed remains ought to be stored within the ever-changing memorial and museum complex.

3

Identifying the Dead

While violence and death on a mass scale is an all-too-common feature of life in much of the world, it was rare enough in pre-9/11 America that most individuals and institutions were not equipped to deal with it. Although the United States had certainly experienced terrorism (e.g., the Oklahoma City bombings) and mass shootings, municipal coroners and medical examiners are trained to deal with death on an individual scale. Bodies are examined, identified, and returned to families in a short time frame. The model is "one body, one death, one family, one time."[1] The World Trade Center attacks not only disrupted families in the New York area and around the world, it also forced the OCME to take charge of a situation unlike anything it had previously handled. For families, there was the massive challenge of dealing with the death of loved ones in a situation that was previously inconceivable. Disease or heart attack, car accidents or house fires: these are the scenarios within our realm of understanding and imagination, not to mention within the medical examiner's realm of familiar experience.

But planes crashing into skyscrapers and skyscrapers collapsing were not. Even in the most horrific accidents, bodies or body parts are retrieved within a few days, examined, and returned to families within a reasonably short period of time. Never before had a medical examiner's office had to deal with a year-long recovery effort, a never-ending identification process, and the formation of deep, long-lasting relationships with families of the victims. Under normal circumstances, a family receives the body soon after the incident and is able to commence the mourning process. But after the World Trade Center attacks, the lengthy process of recovering and identifying the victims meant that families who ultimately received remains of their loved ones had to wait months or years—all while being subjected to an endless barrage of media coverage of the event that irrevocably changed their lives. Shiya Ribowsky, a medicolegal investigator who was the OCME's director of identifications at the time, notes that the waiting times kept many families "spinning in an endless circle of grief" and put the OCME staff who interacted with them under significant stress.[2] This situation upended well-developed OCME routines for managing death at the administrative and psychological level.

It is important to remember that much of the team responsible for identifying the victims was actually at the World Trade Center scene trying to get a sense of the number of bodies they would receive when the first tower collapsed. In addition to their professional roles, they were also survivors who sustained both physical injuries and mental trauma on September 11. This meant that their task at the OCME was not just work—they realized that, but for luck, they may have been among the bodies being delivered to the medical examiner for identification. Thus, the work was intensely personal to them. They bonded with families and felt a duty to victims to ensure that their remains could be returned for proper and respectful burial. According to Ribowsky, "It became difficult to keep a professional shell in place when the decedent's family member that you're talking to became a dear friend."[3]

OCME Processing

What happened to remains, which ranged from nearly complete bodies in the initial hours after the attacks to bits and chunks of charred bone and flesh later on, when they arrived at the OCME from the Pit or Fresh Kills? Assuming remains were not identified as a member of one of the uniformed services, they were placed in a refrigerated truck (initially there were two, one provided by a vendor from the Fulton Fish Market, and eventually there were twenty-four) upon arrival at the OCME intake area on 30th Street. As soon as possible, they were triaged by a forensic anthropologist—either Amy Mundorff or members of the Disaster Mortuary Operational Response Team (DMORT), the network of experts in mortuary sciences and identification techniques that is activated to respond to mass disasters by the federal government.[4] Remains determined to be those of a member of the uniformed services were usually brought immediately into the OCME for triage. FDNY personnel were zealous about the remains of their brethren, requesting special attention for the remains of firefighters during the triage process. While some OCME staffers bristled at this (which they referred to as "Salute the Boot," because that was often all that could be found), others, including Ribowsky, eventually accepted it as a reasonable request from a tight-knit community that put its life on the line to keep the public safe.[5]

Especially in the early stages, the first step of triage meant ensuring that the remains were indeed human. At this stage, workers were cautious and sent for analysis any material that could conceivably have been human remains, including meat and bones from one of the many restaurants at the site, chunks of pulverized concrete that were often indistinguishable from damaged bone to all but the most highly trained specialist, as well as things that were downright bizarre, including, according to Ribowsky, "a set of plastic novelty crooked teeth, yellow and gnarly, clearly from the desk of some office prankster."[6]

Forensic anthropologists had another crucial job: to tease apart comingled remains. As Mundorff and her colleagues discovered, tissue from two or more people was often found together when body bags from the World Trade Center site were first opened. Tissue from one person might be found clinging to the tissue of another, and body parts that recovery workers thought to be from the same person might have been placed together even though they were in fact from different victims. In many cases, parts of one or more victims were found embedded within the body cavity of another victim. In one particularly gruesome case, the OCME found a hand and an amputated finger in a chest cavity they x-rayed. Both the hand and the finger had wedding rings on them, but neither turned out to belong to the victim. Thus, within this one torso, there was biological material from at least three different victims of the 9/11 attacks.[7] Further, especially as the recovery effort progressed, body bags contained numerous small- and medium-size bone fragments that belonged to several victims.[8] In an interview, Mundorff told me that they were so concerned about comingling that even if a set of ribs were found together, if they were not connected by tissue, they were separated and became unique cases.[9]

Since the goal was to ensure that as many individual victims as possible were identified, whenever fragmented remains were not physically attached by hard or soft tissue, or could not be fit together with certainty, they became separate cases. Even when it appeared that various fragments belonged together, but there was no anatomic proof of this relationship, they were separated. The idea was that they would eventually be reassociated through DNA testing if they were indeed from the same individual. The concern about comingling became so pronounced that the OCME ultimately hired three anthropologists to reexamine thousands of remains processed in the early weeks of the investigation to make sure that they were clearly single cases and not from multiple victims.[10]

Once remains underwent a preliminary examination by pathologists, they were sent through a series of forensic stations (photography, fingerprinting, odontology, radiology, etc.), where scientists and lab workers examined and analyzed remains for identifying features (e.g., tattoos, jewelry and other personal property, clothing), gathered evidence that could be used for investigatory purposes, and took tissue or bone samples to be used in the DNA identification process. Each remain received a unique identification number preceded by the prefix DM01—short for "Disaster Manhattan 2001."

Uniformed service personnel—both local and national—assisted at all stages of the triage process. New York and New Jersey Department of Corrections officers not only transferred remains from delivery trucks to the refrigerated trailers that had been set up on site, they also accompanied each individual body or body part through the various stages of examination.[11] Later, an OCME forensic pathologist performed a final examination and issued a death certificate if there was enough evidence to do so. The cause and manner of death were almost always the same: blunt force trauma, homicide.[12]

After remains made it through all stations, they were sent down the street to a location behind the old psychiatric wing of Bellevue Hospital that eventually became known as Memorial Park—a facility consisting at its peak of twenty-four refrigerated trailers, and then stand-alone refrigeration units, respectfully enclosed by a large tent replete with American flags and other patriotic touches. For families of the missing, Memorial Park was a special place that allowed them to feel close to their loved ones even in the absence of a body—relatives and friends turned the area into a shrine. They hung photographs of the missing and left notes and meaningful objects there.

The areas around the OCME facility and Memorial Park also became focal points for rituals surrounding the dead when they could not be carried out in the typical fashion. The Jewish community, including numerous volunteers from Yeshiva University's Stern College for Women, took part in the practice of *shmira*, keeping watch over and reciting prayers for the Jewish victims whose remains could not be buried in the customary twenty-four- to forty-eight-hour window in which Jewish law dictates that bodies be buried. Normally, *shmira* is practiced from the time of death until burial, a period in which the body is not supposed to be left alone, but the ritual went on twenty-four hours a day, seven days a week for nearly seven months at Memorial Park.[13] Similarly, Reverend Charles Flood, an Episcopal priest and medical doctor who played an integral role in the introduction of hospice care into the United States and who devoted much of his time and energy to end-of-life issues, conducted nondenominational Friday afternoon memorial services for families (especially those of civilians) and OCME employees.[14] Initially these took place in a tent near the OCME facility, and eventually the services were moved to Memorial Park.

Families appreciated the dignified manner in which the remains were being stored by the OCME. A few even reported that they felt the presence of their loved ones within the trailers themselves.[15] Recognizing its significance, the city, in cooperation with private donors, renovated the site in the fall of 2002. They added a nondenominational worship hall, a waterfall to reduce the intrusion of city noise, and other aesthetic touches to soften the space.[16] The site was refurbished again in 2006—further recognition that despite its temporary nature, Memorial Park still held great meaning for victims' families.[17]

The Identification Effort

Despite its commitment to respectfully storing victims' remains for as long as necessary, the OCME's goal was always to return as many as possible to families for proper burial. The circumstances of the World Trade Center attacks would make this an extraordinary forensic challenge. Looking back on the endeavor shortly after he left the OCME, Ribowsky described it as akin to "building an airplane, in flight, at night, during a storm, with no cockpit light, and only a badly translated set of instructions to guide us."[18]

The sheer quantity of samples was the first problem—nearly 22,000 human remains were eventually recovered, plus innumerable remains that were determined to be of nonhuman origin. The conditions at the World Trade Center site were also a problem: raging fires, the constant presence of water from rain and firefighters, the decomposition of remains over the eight-month recovery effort, and the recovery process all damaged human remains, often to the point at which DNA could no longer be extracted. But the biggest initial challenge was that investigators did not know exactly who had perished when the towers fell. When a plane crashes, airline companies have a verified list of people on the plane when it took off. While bodies might be badly damaged, there is a closed population to which the parts could belong. In the case of the World Trade Center investigation, the population of victims was unknown and the bodies were highly fragmented. This meant that each body part had to be treated as a unique case, and that investigators would have to build a manifest of victims (to whom the remains could belong) as they went along.

It ultimately took the OCME more than two years to produce what can be considered a complete manifest, reducing the number of dead from estimates of around 6,000 in mid-October 2001 to 2,749 by the summer of 2002.[19] This number has been modified slightly as new victims have been discovered over the years, standing at 2,753 on July 1, 2013.[20] It is almost certainly an undercount of the true death toll because in most open manifest disasters, not all victims are reported missing by friends or family, either because no one knows they were there or because of fear of the consequences of making the report. Particularly in the case of undocumented food service and janitorial workers, many families feared deportation so they didn't report their loved one missing.[21] The OCME worked through the UN and the countries thought most likely to be the places where undocumented employees would come from, but had only modest success in producing an accounting of this population.[22]

At its peak, the OCME's missing persons database contained more than 60,000 records. OCME officials knew that many of these reports were from people who were actually alive or were multiple records from the same victim, but they had to go through the painstaking process of "de-duplicating" their database to ensure that each missing person had only one entry. Consider, for example that four people—a victim's wife, his mother, a friend, and a coworker—all filed a report for a victim we will call Robert Jones. The victim's wife and mother both reported him missing as "Robert Steven Jones" and got his basic vital information correct (e.g., date of birth, home address, employer). But Robert's mother reported his full middle name, while his wife provided only his middle initial. Robert's friend knew him as "Robby Jones" and reported him missing this way. He did not include a middle name or an initial. Given that he was a close friend, he got all of the other demographic information he reported correct, but wasn't able to answer all of the questions posed on the questionnaire and left several blank. Robert's coworker also knew him by his nickname, but she spelled it "Robbie" when reporting him missing. She also misremembered his age and reported his year of birth incorrectly. In this case, the OCME would have to go through all four records to determine whether or not they referred to the same person.

Even completely accurate reports could create challenges. In one case, a single person filed eleven missing persons reports on his brother with various law enforcement and nongovernmental agencies (both in his hometown and in New York City) that were sufficiently different to merit separate entries.[23] Add to this the fact that a number of victims shared last names and sometimes even full names. Once they were done de-duplicating the database, the OCME also had to investigate whether people who were reported missing were actually missing, or whether they were just impossible to reach in the early hours of the crisis. Finally, as one would expect in any situation like this one, there were a few cases of fraudulent missing persons claims that would take a tremendous amount of time and energy to assess.

The Importance of DNA

While OCME staff members were busy compiling as complete a list of victims as possible, OCME scientists were able to identify a small subset of victims through traditionally accepted forensic techniques, especially fingerprinting and odontology (dental examination). They also used unique identifying markers like tattoos and congenital bodily anomalies detected through visual inspection or x-rays in a few identifications. These techniques would ultimately prove to be of limited value, however, because of the fragmented and damaged condition of most remains.

OCME scientists quickly realized they would have to rely on DNA profiling, a technique that had only gained widespread acceptance in the American legal system a few years earlier, and that had just recently been developed to the point that it could be used in mass disasters.[24] The effort to identify the victims of the attacks would ultimately become the largest forensic investigation in American history and would play a role in transforming the science of DNA profiling. It led to innovations in approaches for extracting DNA from damaged bone and statistical analysis of partial profiles. It also expanded the scale at which mass identifications could be accomplished. A story in the *Los Angeles Times* went so far as to argue that the OCME and its private and public partners were "virtually reinventing the science of identification."[25] The claim was a bit exaggerated; work had been ongoing in the field of ancient DNA for many years, and many of the techniques and ideas used by the OCME were pioneered, or at least conceived of, in the context of previous disaster investigations, including the U.S. military's work identifying the remains of MIA/POW soldiers, and efforts to identify the victims of the war and genocide in the Balkans from 1991 to 1999. But it was true that the World Trade Center was forcing scientists to develop new biological and statistical methodologies.

The OCME's Department of Forensic Biology laboratory, where DNA testing was carried out, would play the biggest role in this endeavor. Although the DNA lab now resides in a modern, standalone facility, in 2001 it was split between two less-than-ideal locations: the sixth floor of the OCME building at the corner of 30th Street and First Avenue and the fourth floor of nearby Bellevue Hospital. In the years leading up to 2001, the Forensic Biology lab, led by Dr. Robert Shaler, typically processed roughly 3,000 cases a year, mostly rapes and murders, producing around 1,200 genetic profiles from those cases.[26] Within the first year of the 9/11 recovery effort, the OCME would be inundated with nearly 20,000 complex cases—many of which had to be tested several times with multiple testing strategies in order to produce useful profiles.

Almost immediately, Shaler realized that the OCME did not have the capacity to produce DNA profiles for all of the remains that would emerge from the ashes of the World Trade Center. Borrowing a strategy used by the Royal Canadian Mounted Police in the identification of the 229 victims of the 1998 Swissair Flight 111 crash, Shaler and his team decided to share the DNA profiling work with multiple labs and then integrate all of the profiles in a central database for matching to direct or family reference samples.[27]

As far as remains were concerned, the OCME took the lead in extracting victim samples from soft tissue remains, while a company called Bode Technology would extract DNA from bones recovered from the World Trade Center debris. The OCME laboratory conducted profiling on some of the DNA it extracted from remains and sent the rest to Myriad Genetics, a private laboratory in Salt Lake City, for processing. Bode, on the other hand, produced profiles for all DNA it extracted from the bone samples. On the reference sample side, the New York State Forensic

Laboratory in Albany extracted DNA from direct reference samples and sent it to Myriad for processing. Myriad was also contracted to extract and profile DNA from familial references samples. Wherever the profiles were created, they were sent to the New York State Forensic Laboratory, which had agreed to create a large database to manage all of this information based on the FBI's CODIS system. OCME personnel would manage the entire process.[28]

Additional partnerships would emerge over time as new techniques were used to maximize the amount of genetic information extracted from each remain as well as the ability to analyze it. These included work with Celera Genomics on mitochondrial DNA typing, the National Institute of Standards and Technology (NIST) and Bode on an STR-based DNA test that was more compatible with highly degraded biological samples, and Orchid Biosciences on a single nucleotide polymorphism (SNP)-based test.[29]

Shaler notes that, in an ideal world, the OCME would have held off beginning DNA testing until all of the processes and systems needed were in place, but OCME staff felt tremendous pressure to start immediately. This pressure was partly internal. OCME staff, especially those who had narrowly escaped death as the towers fell, felt the need to do something to help families of the missing. But there was also a public expectation, whether legitimate or not, that DNA would be the key to resolving the fate of them missing.[30] This expectation worried Shaler and his staff because they knew that even the best DNA profiling technology could not identify tissue that contained no recoverable traces of DNA. Yet Shaler believed that "because of DNA's heralded success in criminal cases, rescuers, politicians, the media, and families had a preconceived idea that DNA was infallible. Frankly, DNA had achieved an aura of mystical proportions."[31] More pragmatically, there was also the reality that the cases were piling up. By the end of the day on Wednesday, September 12, the Department of Forensic Biology had received 91 biological samples; by Thursday the number was 266; by Friday, September 14, the total was 639; and by early October, the total was well over 3,000.

How DNA Identification Works

DNA identification is based on the fundamental premise that, with the exception of identical twins, no two human beings have exactly the same genetic make-up.[32] Although all humans share a nearly identical set of the three billion "letters" that make up our genome, each of us also carries with us a substantial number of mutations (i.e., changes to our genetic material) that make our DNA profile unique. Indeed, any two randomly selected individuals will have different letters at about one in one thousand sites. Some of these changes are harmful or even lethal (e.g., in sickle-cell anemia, Huntington 's disease, Tay - Sachs disease, and cystic fibrosis, to name a few well-studied genetic diseases), while others either have no effect, or confer some minute advantage or disadvantage.

The human body contains approximately one trillion cells, all of which are descended from a single fertilized egg. The genetic material, DNA, is found in the form of chromosomes of the innermost part of the cell, the nucleus. The fertilized human egg has twenty-three pairs of chromosomes, with one copy coming from the mother and the other coming from the father at the time of conception. During the course of embryonic development, and cell division throughout life, these chromosomes are completely replicated numerous times. One result of this process is that all cells from a single person, be they hair, skin, blood, semen, or muscle, contain the exact same DNA sequence.

Chemically speaking, DNA is a repetitive polymer of four different types of nucleotides: adenine (generally abbreviated as "A"), cytosine (C), guanine (G), and thymine (T). Under normal circumstances, DNA is composed of two strands that are linked together by bonds that form between the nucleotide of each strand according to a simple set of rules: cytosine pairs with guanine and adenine pairs with thymine. For example:

A T T C G G A A C T
T A A G C C T T G A

Each linked dyad (either A-T or C-G) is called a base pair, or bp for short, and this is the unit of measurement for DNA. The DNA section above is a 10 bp fragment. Long DNA fragments are measured by the thousands, or kilobases. Thus a 7,470 bp long fragment would be written as 7.47 kb.

DNA has many functions. The sections typically called "genes" contain the code for producing proteins and organizing them into cells, organs, and body parts. Other parts of the genome are necessary for the structural integrity of the chromosome. Still other parts have no clear function and are often called "noncoding DNA." This is where most of the variation used for DNA profiling is found. Within these noncoding regions of the genome, there are places where short sequences of DNA are repeated one after the other. Although scientists do not understand why these repeats occur, it has been shown that the number of repetitive sequences at certain locations in the genome is highly variable from person to person. As a result, they are called "short tandem repeats," or STRs. Each fragment of a different length is called an "allele." This term is used widely in genetics to refer to the variants found at a particular physical location, or "locus," in the genome. At any given locus where STRs occur, there can be numerous alleles, each with a different number of repetitive units, and therefore a different length. The presence or absence of particular alleles is what makes individuals identifiable through their DNA profiles.

STRs are visualized using a technique called polymerase chain reaction (PCR), an enzymatic process that replicates a specific region of the genome over and over again. The process entails heating and cooling DNA at precise temperatures through about thirty cycles in the presence of various chemicals, enzymes, and molecules. The most important components of the PCR mix are the two "primers," or short nucleotide sequences that are complementary to DNA sequences at either boundary of the region of interest. These primers bind to the region of the genome being targeted for amplification and serve to attract the thermo-stable enzyme that actually replicates DNA. These primers are labeled with a colored dye, which is used to measure the length of each locus when the results of the PCR reaction are analyzed. Called *Taq* polymerase, this enzyme uses the nucleotides (A, C, T, and G) that are also present in the reaction mixture to copy the DNA between the two primers. In addition to these components, various salts and water are also added to the mixture to stabilize the reaction.

The first step of the PCR process involves denaturing the DNA, or separating it into single-stranded molecules. This is done by heating the DNA to a high temperature, approximately 94 degrees C. Once separated, the reaction mixture is cooled to approximately 60 degrees C, when the primers bind to the DNA sequences to which they are complementary. Next, the mixture is heated to approximately 72 degrees C, when *Taq* polymerase locates the primers and copies the region of DNA between them, forming a double-stranded copy of the original DNA molecule. Once this process is complete, the temperature is once again raised to 94 degrees C and the cycle, which takes a few minutes to complete, is repeated twenty-nine more times.

At the end of the reiterative PCR cycle, several hundred million copies of the regions of interest are produced from only a few original DNA molecules.

The next step in the process is to separate the PCR products by running them through a small capillary tube subjected to an electrical current containing a viscous material through which the DNA fragments pass. This process is almost completely automated and is carried out by specialized machines. The smaller fragments travel faster than the longer fragments and therefore move more quickly through the capillary tube. As the fragments get close to the end of the tube, they pass through a laser beam, which causes the dye bound to fluoresce at the wavelength, and therefore a unique color. The machine's camera records the time, color, and intensity of light when each DNA fragment passes through this "detection window." This information is recorded as a series of peaks on an "electropherogram" (which looks a bit like a Technicolor heart-rate monitor), and a specialized computer program interprets this data. In the final step, data from the machines is imported into a computer program that identifies individual alleles.

It is important to remember that the computer program must distinguish between the numerous alleles of more than a dozen STR loci from data that is not always perfect. Numerous technical problems can lead to interpretation mishaps and occasional misidentifications, most notably "allelic dropout," in which an allele at a given locus cannot be clearly determined. Currently, there are dozens of STR markers available for use, but the FBI has chosen thirteen of them for its DNA database and these were the markers originally deployed in the World Trade Center investigation.

Making a Match

There are two ways of identifying the source of human remains using DNA. The easiest is direct matching: matching DNA profiles from remains and those from a sample known to come directly from a victim. Such samples could include past tissue or blood samples taken for medical procedures or bone marrow donation (firefighters participated in a bone marrow donation program and therefore many had blood samples on file), saliva from a toothbrush, hair from a hairbrush, bodily fluid from previously worn underwear, or skin cells from a razor. The primary challenge with direct samples is that it is very difficult to demonstrate conclusively that DNA extracted from any of these sources (other than blood from a medical facility) definitely comes from the victim. All are ripe for contamination and, other than blood or tissue samples, none provide pristine biological samples for DNA extraction. In the case of the World Trade Center investigation, clean, uncontaminated direct samples did not exist for most victims.

When direct matching cannot be used, investigators turn to kinship analysis. This method is based on the statistical principles of genetic inheritance and requires DNA profiles of blood relatives of the victim. Scientists compare unknown remains to relatives of the missing who have donated "reference samples." Kinship mapping is only straightforward when both biological parents are able to provide DNA samples. Because a child inherits half of his DNA from each parent, this situation usually offers scientists all the information they need to make a match since they can trace all alleles back to their source. Even in this best-case scenario, however, complications can arise that make analysis difficult. If the remain in question generates only a partial profile, which was quite common among 9/11 victims, it is hard to determine the exact familial relationship between the victim and a reference sample donor—parent-child, aunt-niece, and so forth. Further, if a child has rare genetic anomalies not present in either parent or if there are test-induced anomalies in the DNA profile (such as allelic dropout), the kind of matches that one would expect between parent and child will not occur. While these latter are fairly rare, they occur often enough to be a challenge in an identification effort of the scope being carried out by the OCME.

In a large number of cases, DNA from both parents is not available due to death, estrangement, adoption, misattributed paternity, or physical distance. In these cases, investigators must collect DNA from more distant relatives in order to get enough genetic information to declare a kinship match with any degree of certainty. Because of the nature of genetic inheritance, many relatives are required to give DNA samples when both parents are not available for testing. Even full siblings share the same genetic profile for a particular marker only 25 percent of the time, and there is a 25 percent chance that they won't share any alleles at all for a given locus. Further, there is a 75 percent chance that first cousins won't share any alleles at a particular locus, and a 50 percent chance that grandparents and grandchildren won't share any alleles at a given locus.

Even if allelic matches are found, there is no guarantee that a familial relationship exists, since most alleles are present in some percentage of the human population. Just as finding two women who are five feet, six inches tall, have curly brown hair, brown eyes, and brown skin does not guarantee that they are related, finding common alleles in two samples does not necessarily mean they are a match. Thus the power of a genetic marker to discriminate among individuals only emerges when it is combined with several others in different parts of the genome. This is precisely why forensic scientists examine many loci before declaring a match, and only do so after calculating the statistical probability that the two samples match because they come from the same source and not by chance.

The OCME and its collaborators had to build an information management system that allowed them to compare all DNA profiles from remains with all DNA profiles donated by family members, calculate how similar or different they were, and then build a probable identity for remains based on how well they fit into particular familial lineages. The OCME ultimately used three programs to maximize the number of matches they could make: a purpose-built matching system called MFYSIS (which was created by a company called Gene Codes Genetics), a program written by mathematician Charles Brenner called DNAView, and a program written by Canadian forensic scientist Benoit Leclair called MDKAP.[33]

Both direct matching and kinship matching were subject to an additional technical challenge: the damaged and degraded nature of the reference samples meant that the OCME and its collaborators often could not get full DNA profiles from remains using the technology available in 2001. As such, they had to figure out how to do more with less, both in terms of developing techniques to extract genetic information out of short, highly degraded samples of DNA and to combine partial profiles extracted using many different techniques into a single "virtual profile." While such efforts were highly unconventional and would not have been considered admissible in a court of law, they offered at least a hope of identifying the remains.[34]

Given all of these complexities, OCME scientists had to be careful when trying to make DNA matches. At first, this was all done manually, which was tedious and stressful for everybody involved. Shaler once had a nightmare that Karen Dooling, the person responsible for leading this matching process, had to move into the OCME laboratory and live there permanently. He writes, "her husband Mike was banging on the door trying desperately to grab her attention, but she was sitting on the floor of the conference room sorting through CODIS match lists. Stacks of candidate match sheets surrounded her, ripped up and wadded-up paper was strewn around, used color highlighters had been tossed in a corner, her ponytail stuck straight up in the air as though she had poked her finger into an electrical socket."[35] Putting the problem more simply, if less vividly, Shaler admitted, "I had no way to convert DNA test results into identifications" in a rapid, automated fashion.[36]

In an effort to develop a system that could make matches in at least a semi-automated fashion, Shaler and his staff invited leading experts from public laboratories and private firms to brainstorm possible solutions to the data analysis problem on October 3, 2001. The media, grasping for positive stories about the World Trade Center attacks, picked up on the meeting as a sign that the United States was at least beginning the process of healing. For Shaler and his OCME team, though, the meeting emerged less from boldness and strategic thinking, than from exasperation. Shaler notes that while the *Wall Street Journal* chose to headline their article on the meeting "Summit Called to Map Strategy on Victims' DNA," a more accurate title would have been "OCME Scientist Desperate for Help."[37]

The October 3 conversation concluded that no existing software package could handle the complexities of matching partial profiles to a large database of familial reference samples. The OCME would have to work with contractors and outside firms to develop one. This endeavor would ultimately be guided by a high-level advisory group put together by Dr. Lisa Forman, director of the Investigative and Forensic Sciences Division at the National Institute of Justice (NIJ) specifically to assist the World Trade Center identification effort: the Kinship and Data Analysis Panel (KADAP). KADAP met for the first time between October 18 and 20, 2001, just one week after the OCME put in an official request for assistance, and then met eight additional times through July 2003. The panel was funded by the NIJ and included more than thirty world-renowned scientists from government, industry, and academia, all of whom had played important roles in the development and implementation of DNA profiling techniques in law enforcement or research laboratories.[38] Among the most notable were Bruce Budowle, who led the development of DNA Identification within the FBI Laboratory; John Butler from the National Institute of Standards and Technology, whose work on the biology and population genetics of STR markers is considered foundational in the field of forensic science; and Kenneth Kidd, a population geneticist who helped the FBI develop their methodology for determining the probability of a false match between a crime scene DNA profile and a suspect's DNA profile. KADAP also had specialists in kinship matching (Brenner and Leclair); extraction of DNA from damaged and degraded tissues (Thomas Parsons from the Armed Forces DNA Identification Laboratory); and the management of large-scale identification projects (Amanda Sozer).

In addition to KADAP's ambitious agenda, Shaler had a more immediate problem for them to solve: he needed statistical cut-offs that would allow him to make manual matches in cases involving partial profiles from remains. At the time, there were no generally accepted guidelines for doing so whether from direct reference samples or familial reference samples.[39] While the criminal justice system required matches at all thirteen loci to declare a match, most of the remains from the World Trade Center site were yielding only partial profiles, when they were yielding anything at all. If OCME scientists followed the accepted standard, their ability to make matches would be dramatically curtailed. They needed to come up with a way to balance scientific rigor and the realities of the work they were engaged in.

After deliberating on the issue, KADAP recommend using a likelihood ratio of 1 in 10 billion to determine the validity of partial matches between remains and direct samples. According to the panel, this was the matching standard that had to be used to guarantee no errors if one wanted to be at least 99 percent certain of the match result. While confusing to explain to the public, this likelihood ratio had a practical translation: it would be statistically valid to declare a match if the DNA profile of a remain matched at least ten of thirteen loci of a direct reference sample. For kinship identifications, the panel recommended a minimum probability of 99.9 percent to declare a match between remains and a family group. Perhaps most importantly, KADAP also recognized the unprecedented nature of the task at hand. It recommended that the OCME exploit "research-grade" systems to produce matches when forensically validated systems could not do the job.[40]

The First Matches

The first DNA matches were made on October 12, and both involved direct samples taken from victims' toothbrushes. These particular matches were cases that had already been resolved through other means, so DNA was not the primary technology used to identify these particular remains, as it would become for the majority of later cases. However, the fact that the DNA result confirmed the identification made through other means suggested that the system that Shaler and his team had put in place, and the tremendous amount of work they put into validating it, seemed to be working.[41]

Over time, the ratio of DNA matches to other modalities shifted dramatically as the percentage of relatively complete bodies or undamaged soft tissue being brought into the OCME declined and the percentage of bones and badly putrefied or damaged soft tissue increased. By the end of October 2001, for instance, only 10 of 193 total identifications were based on DNA. At the end of March 2002, 208 of 862 total identifications involved DNA. By June 2002, 495 out of 1,203 identifications involved DNA. At the one-year anniversary, 673 of 1,496 total identifications were based on DNA. This pattern would hold for the rest of the project.[42] By July 2013, DNA was responsible for 886 of 1,634 total victim identifications and was used in 574 cases in conjunction with other forms of evidence.

The OCME's identification of the hijackers' remains was entirely reliant on DNA profiles recovered by the FBI from objects left behind in hotel rooms and rental cars in the days leading up to the attacks. During the course of the investigation, the OCME has identified remains of four of the ten hijackers involved in the attacks on the World Trade Centers through DNA, and is currently storing those remains in an undisclosed site sequestered from remains of their victims.[43] This decision was made both out of respect for the victims and their families, and the view among OCME officials that it is simply the right thing to do.[44]

The next most important technique, odontology, was responsible for only 49 identifications alone and was involved in 482 identifications involving more than one technique. Fingerprints were responsible for 33 sole identifications and were involved in 272 cases involving multiple methods.[45] This breakdown makes sense given that dental examination depends on the recovery of reasonably complete jaws with intact teeth, and fingerprinting relies on the discovery of reasonably undamaged fingers to produce identifications. Not many of either were recovered.

Family Involvement

In addition to resolving a number of incredibly challenging technical and organizational issues, OCME staff also had to navigate the complexities of relatives' emotions when carrying out the identification effort. Sometimes families were angered by or frustrated with city officials. Other times, they remained quietly despondent, out of the view of cameras and reporters. And still other times, they channeled their feelings into advocacy and change. Either way, the relationships that family members and OCME staff developed fueled early efforts to get the identification effort underway and blurred the boundaries between OCME scientists and the public they were serving. "These wonderful people," Shaler writes in his memoir, "bonded with me and my staff, made us feel wanted and important, consoled us and showed us why our efforts were worthwhile. . . . We often cried at our computers and at weekly meetings with the families. Sometimes we high-fived after identifying someone or we'd steal away to an out-of-the-way corner to be alone or simply go for a walk to deal with our emotions. And we hugged."[46]

One example of a positive outcome emerging from grief is the story of Give Your Voice, which was founded by the family of twenty-six-year-old James Marcel Cartier, an apprentice electrician who had been assigned to the World Trade Center in August 2001. Within a few weeks of September 11, James's sister, Jennie Farrell, his father, Patrick Cartier, his brother Michael, and other family members became spokespeople for civilian families—advocating for what they perceived to be more equal treatment of their loved ones' remains and better communication between these families and city officials on the recovery and identification effort.

By early December Give Your Voice was beginning to ask questions both publically and privately about the identification effort being carried out by the OCME. Families were confused about the process and could not figure out whether the biological samples they had given to the NYPD at Family Assistance Centers were ever going to lead to the return of their loved ones remains.[47] In their view, the process was taking too long and city officials were ignoring their questions.[48] They decided to take matters into their own hands and contact the OCME directly.

Shaler recounts a phone conversation with Patrick Cartier in early December 2001, when Patrick called to inquire about the status of the sample that he had provided the NYPD at Pier 94. Shaler remembers the anguish in Patrick's voice and his desire to know that everything that could be done to identify his son's remains was being done. When Shaler called members of his staff about the case, he was startled to learn that, while Patrick's full DNA profile was indeed in New York State's World Trade Center database, no other DNA samples had been taken from Patrick's relatives.[49] A kinship match could not be made without a maternal sample, or additional close relatives. This oversight was not the result of a lack of reference donors, but an apparent failure of the NYPD to request appropriate samples.

Conversations like this confirmed what OCME staff were already beginning to suspect—that there were serious problems with the antemortem data and familial reference samples collected by NYPD and DMORT at the Family Assistance Center.[50] First, despite having the best intentions, many of the officers and DMORT personnel who staffed the FAC made clerical errors when collecting and entering information about victims from family members. In an analysis done a year after the attacks, Mike Hennesey found that nearly one in six records (out of 4,500 total) had either inaccurate data or recording errors, ranging from trivial (mistakes in spelling of names or dates of birth) to more serious (incorrect familial relationships recorded). Second, hundreds of cases could not be solved because of insufficient reference and victim DNA samples.[51] There were several reasons for this problem: the sample may have been lost or mislabeled; the cheek swab may have been incorrectly obtained; or a laboratory processing error may have occurred. In other words, FAC personnel did not always fully grasp of the scientific requirements of DNA identification process and either did not collect—or mishandled—familial samples, or collected direct samples that could not be uniquely and conclusively tied to a victim.[52] This was a problem faced by the Cartier family, among many others. Although James's remains had been discovered and were being stored by the OMCE, they did not have sufficient kinship samples to make a scientifically valid match because only Patrick had provided a sample.

More generally, families were being overwhelmed by the complexity of the system, their inability to get information about it, and the barrage of reference numbers they received each time they interacted with city officials. When a relative or friend came in person into the FAC to report a loved one missing, they received a "P" number. If they called the NYPD to report a missing person, they received a "T" number. Further, each time a family member brought in a direct sample, or gave a kinship sample, they received a DNA collection number. Additionally, in the first few days after the attacks, families often brought direct reference samples to the OCME for collection. When this happened, the OCME assigned the family a family reference (FR) number. The number of numbers a particular family would have was only limited by the number of times they interacted with the system. As Shaler notes, the Cartier family had "two P numbers, a T number, an FR number, and five DNA collection numbers. Multiply this single family by thousands, and the resulting confusion was mind-numbing."[53] Add to this the additional numbers generated once the samples were processed by the laboratories and entered into the New York State Police system, and it becomes clear that the WTC investigation was a numerical nightmare. Although this problem was never fully solved, the

OCME eventually succeeded in bringing most of the numbers from a given case together under a single reported missing (RM) number.

Throughout December 2001, OCME officials had numerous discussions with family members to try to understand their needs and determine how better to serve them. They recognized that this would increase the information they had access to and would make identifications easier. They also gave several family members tours to explain how the identification operation worked. At the December 29 meeting of Give Your Voice, Michael Cartier reported that he was extremely impressed with the entire recovery operation and found the staff at the OCME to be competent and kind.[54] Visits like this helped both sides: families wanted to know more about the process, especially how long it would take, and the OCME needed adequate samples and correct antemortem data.[55]

After consulting with family representatives, the OCME did two things: first they decided to create an information booklet for families that detailed the DNA identification procedure and provided answers to commonly asked questions. More importantly, they also opened a phone bank on January 26, 2002 that relatives could use to check on the status of their case, ask questions, and find out whether the OCME had enough familial samples to make an identification in the event that DNA was recovered from a loved one's remains. The hotline was staffed by medicolegal examiners with knowledge of DNA identification science. It received a huge volume of calls—more than 1,600 in the first two weeks of operation—and resulted in more than 700 follow-up appointments to collect additional DNA samples.[56]

The OCME was honest with family members when answering their questions. They did not hide the fact that there were numerous problems with the samples collected at the FAC and that there was a good possibility that families would have to resubmit samples or submit samples from additional family members.[57] These new samples would replace improperly collected familial cheek swabs and direct samples that had yielded no DNA, and would also provide DNA from the additional family members needed to make kinship matches. The outreach was so successful that by July 2002, the OCME reported to Give Your Voice that fewer than fifty cases were outstanding because of the lack of available reference samples.[58]

Having to search for direct samples or to provide more kinship samples proved difficult for many families. In a *New York Times* story about efforts to locate new samples, relatives of the missing described how they were affected. Stacey Staub, whose husband, Craig, was a trader at Keefe,

Bruyette & Woods, "turned in a toothbrush, a bag of hair, a toenail clipper, dirty facial tissues and the dirtiest underwear she could find" in the early days of the frantic search for victims. But she could not bring herself to touch his other possessions after that initial submission. She left his towels in place and the glass that he drank from the night before his death on his nightstand. She was upset when she found out that the direct samples she provided to police yielded no usable DNA—the call for more material was for her like "opening the wound that's just been sewn over and over and over again."[59] Pamela Block Works, whose husband John was a colleague of Craig Staub, expressed similar sentiments, telling the *Times* reporter: "I want to do anything that can help them make some sort of a positive ID. But sometimes, you just wonder whether this is going to become an exercise in futility."[60]

OCME staff also did not hide the fact that even with adequate samples, identifications were not guaranteed. The terrible conditions at the World Trade Center site meant that more than 60 percent of remains recovered, especially those pulled out after the first few weeks, provided either partial STR profiles or no profile at all.[61] Even when identifications were made, families did not automatically find relief from grief. The OCME explained in its information booklet that "some families may find comfort in knowing that the remains of their loved one have been identified and returned. These remains can be interred according to the family's traditions. This may help with the healing and adjustment to this terrible loss. For others, the DNA testing process may interfere with their healing. For DNA testing to work, we may need more information, DNA samples, or personal effects. Gathering these may cause you further distress. If DNA testing does not identify your loved one's remains, it may be a disappointment, adding to your grief. These issues are very personal and may be different for each family."[62]

More generally, OCME staff decided to be completely transparent about the mistakes it made during the identification effort. They understood that the only way to build trust with families—who were growing increasingly skeptical of the city's commitment to helping them—was to be as honest as possible in all facets of their interactions, especially when they could just as easily have remained quiet and no one would have been the wiser.[63] The hardest mistakes to admit were those that involved returning the wrong remains to families. This happened on a few occasions.

The best-known case involved the misidentification of a New York firefighter that ultimately became the basis of a front-page *New York Times* story.[64] Firefighters working at the pile recovered the remains of a colleague that they insisted was Jose Guadalupe from Engine Company 54. He was found exactly where they expected him to be, based on radio transmissions and their knowledge of NYFD activities just before the collapse. Guadalupe was known to have a rare spinal anomaly in one of his vertebrae that only occurs in 1 in 100,000 people. Sure enough, when the OCME x-rayed the remains, they found the anomaly—making the identification seem like a slam-dunk. What were the odds that remains found exactly where a firefighter was expected to be, wearing jewelry that he was known to wear, and having a very rare skeletal anomaly was not the firefighter? The OCME assumed they were low and handed over the body to the Guadalupe family for reburial.

It turned out that Guadalupe had an older mentor at his firehouse named Christopher Santora. Guadalupe looked up to Santora and mimicked his professional behavior and his personal style. He wore the same jewelry as Santora and had his hair cut the same way. What they possible could not have known while they were alive is that they shared the same rare skeletal anomaly. When DNA was analyzed from what was thought to be Guadalupe's body before it was returned to his family for reburial, OCME scientists made a startling discovery. It did not match his direct samples or any of his relatives who donated DNA in the early days of the recovery effort. Rather, it matched DNA taken from Santora's toothbrush. This meant that the OCME would have to break the news to the Guadalupe family that the remains returned to them—and now buried—did not actually belong to Jose.

When Shiya Ribowsky explained this situation to Gualdalupe's widow, she broke down. Years later, Ribowsky recounted, "I am unable to adequately describe to you how painful this was for her. I wasn't sure if she was grieving more for herself or for the pain the Santora family was about to experience."[65] Santora's family was equally upset by the discovery. His mother, Maureen Santora, was quoted in the *New York Times* as saying, " 'It's hard to tell in words what my reaction is. My first reaction is, how terrible it is for the Guadalupe family. I feel awful this is on the back of someone who had closure. I'm nearly speechless. I can't explain how this happened. I can't be angry. I've been praying all these weeks and crying and not knowing whether they would find a body. I can't be angry. But this mistake has caused a lot of grief."[66]

Several other mistakes were made in the early weeks of the investigation. In one case, a torso was handed over to a family based on an ID card that was found in its shirt pocket, but subsequent DNA testing revealed that the card did not match the torso. Although no one knows how the ID card ended up there, the best guess is that the victim grabbed the ID of another person who had already died in the attacks in order to report his death when he got out of the building.[67] Another situation occurred when two left arms were returned to a single family because of a mistake in the storage of familial reference samples during the collection process. Because of similar P numbers, the reference samples from one family were accidently placed in the same bag as the reference samples from another family, making it seem as if the two families were one.[68]

Even the family members who became most critical of the OCME and their handling of remains after 2005 told me they were grateful both for the work and honesty of OCME staff during the first few years of the identification effort. Diane and Kurt Horning, who cofounded World Trade Center Families for Proper Burial, were notified four times about the discovery of additional remains after they received the first set of remains of their son Matthew. Diane also received the call that no family wants to get—that some of the remains that were handed over to her were not Matthew's. While the Hornings were traumatized by having to give back remains that were being held by a funeral home for eventual burial, they admired OCME staff for their honesty and commitment to truth above all else.[69] Thus, even while they were engaged in a bitter fight with the city over the storage of remains at Fresh Kills, they maintained a good relationship with the OCME staff.[70]

Relatives responded well to this matter-of-fact approach and worked with the OCME to improve their processes and outcomes even in the face of tremendous frustration about the pace of identifications. According to Mundorff, families who met with the OCME either individually or in group meetings provided critical feedback on current and future policies and practices, and became ambassadors to families that either could not, or did not want to engage with the medical examiner. They could explain to less engaged families that the process was taking long not because of incompetence or indifference but because of the legitimate technical and data-related challenges and the need to rigorously validate systems before using them to make identifications.[71]

After the Identification

By early spring 2002, the hard work undertaken in the months after September 11 was beginning to pay off. Identifications were being made on a regular basis and the several missteps and mistakes made by OCME personnel had been discovered and worked out. Additionally, usable versions of matching software were coming online. Families were also beginning to receive remains of their loved ones in larger numbers than in the previous six months.

As a matter of policy, all notices of identification were made in person through police departments. This was seen to be the most effective and humane way to relay the news. Once a match was made and verified by anthropological and administrative review, and a death certificate was issued, the OCME would notify the NYPD. If the next of kin was in its jurisdiction, NYPD would send an officer to the house to explain to the family that their loved one had been identified. If the next of kin lived outside of New York, NYPD officials would work with local law enforcement agencies to send a representative to the family. Police officials would give the next of kin the OCME phone number so they could get further information and arrange for the handover of remains.[72] Families then contacted the OCME directly and could come in to find out more about their loved ones' remains or simply arrange for the handover. All handovers were carried out through funeral homes because they had experience dealing with human remains and grieving families.

Families could not ignore the realities of the World Trade Center site even when they received the remains of their loved one. During the handover process, next of kin had to sign a release form letting the OCME know what it should do in the event that additional remains were discovered. The family could choose not to be notified of any additional discoveries and ask that the OCME store or dispose of the remains "by methods deemed appropriate"; they could be notified each time a new remain was identified; or they could choose to be notified "one time after the medical examiner deems it unlikely that further tissue will be identified."[73] As with other aspects of the recovery effort, there was no standard family response to this question. Different families chose different options for a variety of reasons. While the OCME of course respected the wishes of the families, many representatives of the agency told me that they sometimes wished they could provide additional information to families that had chosen not to be notified.

For many families, the death certificate that accompanied their loved one's remains replaced the special one that they received in the absence of identified remains. It is important to understand that violent death brings with it not just emotional and psychosocial burdens—it also creates a host of practical challenges for families. Estates need to be settled, property needs to be redistributed, death benefits from insurance policies need to be obtained, and aid from government agencies and nongovernmental organizations must be accessed. Death certificates are required for many of these. The process of issuing a death certificate is relatively straightforward and is carried out by the OCME—assuming that there is a body present. When a person goes missing, but is presumed dead, the process becomes much more complicated and requires a three-year waiting period before a death certificate can be issued.

The World Trade Center attacks, however, created a large volume of missing persons cases in which there was little doubt about the fate of victims. The OCME recognized that many families would have to wait months or years for remains, and that a significant percentage would never receive the remains of their loved ones. Under these circumstances, the OCME, the city, and the state determined that a different approach needed to be taken to issuing death certificates for September 11 victims. City and state agencies worked together to develop a process that would provide death certificates to families within a few weeks of their request rather than after the usual three years.[74]

The special process was handled by the New York City Law Department in conjunction with the New York County Supreme Court. Families that wanted to participate worked with Family Assistance Center and Law Department personnel to fill out the required paperwork and affidavits to support their claim. In addition to providing basic information about the missing person, including their address, close relations, education, and employment, next of kin who wanted a death certificate also had to provide details of their last contact with the missing person; why they believed the person was at the World Trade Center at the time of the attacks (i.e., work, an appointment there, commuting patterns, etc.); and a description of efforts taken to locate the missing person.[75]

Fraudulent claims were a concern for city officials, so they required that uniformed service agencies involved in the rescue effort, businesses that had offices or employees at the World Trade Center, and the airlines involved provide supporting affidavits attesting to the missing person's presence at the scene at the time of the attacks. In such cases, the determination of death was little more than a formality and death certificates could be quickly issued. Things got more complicated for families whose cases could not be supported by these methods. The official policy for issuing missing persons death certificates noted that, in these cases, "an individual with personal knowledge of the missing person's whereabouts on that morning will be asked to provide an affidavit outlining why it is believed that the individual was present there on that day. The person filing the affidavit should be prepared to explain when and under what circumstances he or she last communicated with the missing individual, and the basis for his or her conclusion that such individual was at the World Trade Center on September 11. The family member will also be asked to detail in an affidavit his or her efforts to locate the missing person since September 11. In these cases, it may be necessary for the Court to hold a hearing on the facts surrounding the individual's disappearance to examine the basis for the belief that the individual was in the World Trade Center and the efforts of family members and others to locate the missing person." Once the court determined that a request for a World Trade Center-related death certificate was legitimate, it asked the OCME to issue the certificate.[76]

Effect on Families

Each family I spoke to had had a different reaction to losing a loved one in the World Trade Center attacks. Beyond the immediate responses to death—sadness, shock, disbelief, and a desire to know what happened (where was she, did he suffer, who was she with when she died, etc.)—the relatives I spoke to, those whose books I have read, and those whose stories I have heard have handled the issue of their loved one's remains in a variety of ways. One relative might be happy to know that his or her loved one was recovered largely intact because it meant that the body could be buried in the traditional manner, while another might be upset at the news because it suggested that the person may have lived longer after the impact and may have suffered more before death. One relative might be happy to recover even a tiny bit of bone, but for another such a discovery would be a permanent reminder of the fate of their loved one. Similarly, while fragmentation of the body might produce horrific images for one relative, it would provide assurance to another that her loved one did not suffer for long. One family might want to be notified each time one of their loved one's remains was identified, while another family might be content with just a single remain, or even none, because their loved ones' fate was already known and understood.

OCME staff believed that these were all considered "normal" responses to horrific news and it was their job to respect the different reactions of the families. They allowed family members to guide conversations about human remains. They would first tell the next of kin that they had identified a body part and then ask if they wanted to know how many, which parts, their condition and size, and so on.[77] It was always important to find out what the relative wanted to know before divulging information, because, as Giovanna Vidoli, who interacted with families from 2004 to 2006 noted, whatever OCME representatives said could never be "unheard" by the family.[78]

Ribowsky reported that some families wanted "reassurance" from the OCME that the identification was "absolutely positive identification—that and no more." Other relatives, however, wanted more information than the OCME customarily provided. They wanted to see, with their own eyes, and in some cases even touch with their own hands, the remains of a loved one. OCME staff generally tried to steer families away from such practices because remains often bore almost no resemblance to what we would recognize as human—a limb, a fragment of bone, or a chunk of desiccated, decomposing tissue. In many cases they showed relatives photographs, or provided "long, graphic descriptions" of the effect of the collapse of the towers and their exposure to heat and moisture as a first step in the process.[79] This often quelled any desire to be physically in the presence of the remains, but not always. In these cases, the OCME made every effort to organize a viewing of the remains at a funeral home upon handover, but did occasionally take people to Memorial Park to view remains.[80] Ribowsky recounts a moment when Michael Cartier "donned a pair of gloves and held and cradled his brother's remains. It was as touching a scene as I've ever witnessed."[81]

Ribowsky writes that when a positive identification had been made, and he was having a meeting with families to discuss what had been found, he would often start by saying, "we're about to have the most difficult conversation you'll ever have."[82] Despite the difficulty of the conversation, current and former OCME staff told me that some families greeted the news that their loved one's remains had been discovered with great sadness, others with relief, and many with both emotions simultaneously. The sadness, of course, came from the finality of the identification—that there is no longer even a glimmer of hope that the victim is still alive—and the relief came from finally putting an end to that nagging, irrational hope that the victim was actually walking the streets of New York or New Jersey with a severe case of amnesia.

A few examples, some from my interviews and others from media reports, serve to highlight the great diversity of responses. Beata Boyarsky, whose brother Gennady was identified days after the attack by the wallet he was carrying, told a *New York Times* reporter that she could not help but ask why her brother was one of the very few victims identified so quickly. She wondered whether it might have been because he was closer to escaping than other victims. Given that her brother's remains were returned before the decision to do routine identity testing on remains, the notion that she may have received the wrong remains also lurked in her mind. While the return of remains answered some of her family's questions, it left others—particularly the question of why Gennady died—painfully unresolved.[83]

For the Valentin family, receiving a small shard of bone from thirty-nine-year-old NYPD officer Santos Valentin, who was identified from DNA extracted from his toothbrush, transformed their experience of grief. "After not hearing anything for so long, we didn't know if we were ever going to get anything, and just to get this news is a miracle," Santos's sister, Denise Valentin, told the *New York Daily News*. "The DNA process is working," she continued. "Even after all this time, it can be done. Whether it's a small amount or a skeleton. . . . It doesn't matter what they find, it's still the person. It's still my brother. We went from not having anything to having a brother all over again."[84] In a follow-up story the next day, the *Daily News* quoted Ms. Valentin as stating that "we feel we have given him a proper Christian burial, and we hope to start the closure process."[85] Donna Regan, whose husband Robert, a firefighter, perished on September 11, said that receiving Robert's remains helped her children accept that their father was not coming back. Even a small bone fragment helped alleviate what she termed the "vanish factor"—the challenge of saying goodbye to a loved one in the absence of any physical remains to mourn over.[86] Ellie Hartz, wife of victim John Hartz, did not anticipate the emotions she experienced in the aftermath of the attacks. "I have never been able to understand why people have been so intent on recovering bodies. Now I understand. It is a basic human need. We are tactile."[87]

One of the most challenging aspects of the World Trade Center situation was that almost all victims were badly fragmented as the towers fell. This meant that they were recovered in multiple parts at various places and times during the cleanup of the site. As such, many victims' remains were discovered several times over the course of the identification effort. Ribowsky told me that families would often want to know when and where the remains of their loved one were recovered. It was difficult to have to show them a map of the site and say that one piece was discovered in one place in mid-November, another was discovered at another part of the site in late December, and a third was discovered at Fresh Kills in February.[88]

Early on in the process, families told the OCME that it was hard to receive remains, rebury them, and then get another calls about the identification of additional remains. In response to this feedback, the OCME began to offer relatives the choice about how and when to be notified of additional identifications.[89] Some families, including that of twenty-eight-year-old firefighter Mark Petrocelli, chose to receive his remains each time a recovery was made and to add these remains to the coffin that they got for him when his first remains were discovered. In essence, Petrocelli has been buried several times. Others, including that of thirty-six-year-old firefighter Robert Evans, decided to wait until the end of the recovery effort to receive additional remains after burying their son in June 2002, when his first remains were identified and returned to the family.[90]

Another option that families could have was to be contacted through an intermediary. Gordon Haberman, whose twenty-five-year-old daughter Andrea worked for Carr Futures in Chicago and had arrived at Carr's World Trade Center office for a business meeting a few minutes before the attacks, decided he wanted to keep receiving notification every time another part of his daughter was discovered in the rubble, but did not want his wife to be aware of each find. He told me he wanted to protect her from the shock and grief that accompanied each notification. As such, he arranged for all notifications to be funneled through an NYPD detective who had lost his brother on 9/11 and whom Haberman had befriended in his early efforts to find out what happened to his daughter. Haberman told me he has been notified fourteen times about the discovery of Andrea's remains, most recently in December 2013. At the conclusion of the initial identification effort in 2005, the Habermans retrieved all of Andrea's remains that had been found to that date and had them cremated. Since then, they have decided that all additional finds will be stored in the OCME repository within the National 9/11 Memorial and Museum at the World Trade Center site.[91] When I asked him why he continues to be notified of the discovery of his daughter's remains even though it causes him and his wife considerable anguish, and they have no intention of retrieving the remains and bringing them back to Wisconsin, he told me that he has coped with the loss of his daughter by engaging in a continual quest for knowledge about the attacks. He seeks not only to learn about the circumstances behind the collapse of the towers and the effort to locate the remains of the victims, but also about the root causes of the attacks. He has studied Islam at length, particularly its fundamentalist varieties, and has become a passionate advocate for understanding terrorism in an effort to prevent it. Beyond that, he told me that being notified of any finds involving Andrea's remains, and of staying apprised of the OCME's work more generally, helps him stay loyal to her as a father. He feels that she deserves to be thought of, and cared for, even in death.[92]

Families of the Unidentified

Forty percent of the victims of the World Trade Center attacks have never been identified, either because their remains have not been recovered or because they do not contain usable DNA. This too has a profound impact on grief and mourning. Among these victims is Harvey Gardner, a thirty-five-year-old computer consultant for General Telecom, which had offices on the 83rd floor of the North Tower. Harvey's brother Anthony Gardner told me that for the first few days after the attacks, he and his family were still hopeful that Harvey had managed to find a way out of the building. After weeks of frantic searching, however, it became clear to the family that he wasn't alive. Anthony told me that the lack of remains has made the grieving process difficult for his family—they had a memorial service at the end of September 2001 that was attended by hundreds of people, but it was still hard for them to accept that he was gone in the absence of a physical trace of his body.[93]

Still, they have found ways to represent his physical presence over the years. Anthony told me that after the September 2001 memorial service, they brought the flowers from the event (which were red, white, and blue, in keeping with the patriotism that surrounded 9/11) to the beach in Spring Lake, New Jersey, where Harvey often rode his bike after work, and threw them into the ocean. He told me the experience was "electrifying"—the waves were crashing onto the shore, there was a lightning storm at sea, and the flowers were being pushed back and forth along the shore by the waves. He described it as almost being Technicolor. And it was through this energy that they *felt* Harvey's presence for the first time since his death.[94]

Eventually, the family realized that they would not likely get any of Harvey's remains for burial, so four years after his death, Harvey's mother purchased a plot at a cemetery in Wall, New Jersey with a traditional tombstone. In the absence of a body, the family elected to purchase a child's casket and fill it with some of Harvey's belongings. They then had a small funeral. The family still has the urns that the city handed out in October 2001 as well as some sentimental items like his beeper and a beaded necklace he was fond of, but none of these can adequately compensate for the lack of a body. Anthony told me he still hopes to get some piece of Harvey's body back, but at the same time he worries that an identification at this stage would cause his mother grief and make her go through the mourning process again.[95]

In the absence of a body, the actual World Trade Center site has taken on tremendous significance to him—he has been actively engaged in debates about the preservation of the remaining remnants of the towers, and he and his family return to the site each year for the anniversary ceremony. "It is something I will always do," he told me. "We all develop traditions and rituals because you don't have tangible remains and that final goodbye."[96]

Not all relatives were interested in the return of remains, however. Beverly Eckert, who was well known first for her activism in the wake of September 11, and then for her untimely death in the February 2009 plane crash of Colgan Air Flight 3407 near Buffalo, New York, publically stated that she did not want the remains of her husband Sean Rooney (who worked at Aon Corporation on the 105th floor of the South Tower) to be identified. She said that receiving remains would force her to "think about what the particular remains found mean to the way he died. . . . I'd prefer, in my mind, to somehow think that there was this total instantaneous disintegration and that his remains haven't been sitting in a refrigerated trailer all this time."[97]

Michael Raguso's parents Vincent and Dee Ragusa held on to the hope that their son's remains would be discovered for more than two years. Finally, in December 2003, the family decided that they "needed to put Michael to rest," so they buried a vial of his blood that was taken when Michael enrolled in FDNY's bone marrow donor program as a new trainee. Ragusa's funeral was the last held for the 343 FDNY firefighters who lost their lives on September 11. His parents decided that the burial of his blood would be end of their waiting period and the beginning of their efforts to come to grips with Michael's death. Indeed, they asked the

OCME not to notify them if any of Michael's remains were found in the future.[98]

The diversity of responses both to the return of remains and their absence highlights one of the truisms of the World Trade Center story: there is no such thing as a single normal reaction to death, especially in the context of an act of atrocity like 9/11. In fact, the public nature of the deaths made grief and mourning all the more complex. Families describe dealing with the deaths of their loved ones as a saga that waxes and wanes but never truly reaches any sort of definitive conclusion.[99] Lynn Castrianno, who lost her thirty-year-old brother Leonard, a Cantor Fitzgerald employee, described the differences between his death and her sister Caroline's, who died suddenly from acute pneumonia in 1996. She told me that her family had already lost a sibling, so they "knew what 'normal' grief felt like." Grief and mourning for Leonard was anything but normal because of the national scope of the 9/11 tragedy. She continued:

> With my sister Caroline, the process was personal. It was mine. I owned it. Nobody but my family knew what it was like to lose *my* sister. With 9/11, to this day, when you mention you lost someone on 9/11 it's everybody's grief. It's national. You don't own it—everybody owns it. Everybody tells you about their feelings and their grief, and where they were on 9/11. Sometimes I don't tell people I lost someone on 9/11 because I don't want to hear about their experiences. People call you on the anniversary to see how you're doing. Well, for me, May 7, the day that Caroline died, is equally significant. But people just don't get this. Why should my brother matter more than my sister? It's cognitive dissonance because I'm like why is Leonard's death more important than Caroline's? They are equally important to me, but everyone wants to hear about Leonard because it's national.[100]

This constant presence of 9/11 in American society, political debate, entertainment, and even casual conversation makes it hard for her to "come to grips" with Leonard's death. "With Caroline," Lynn told me, "she was alive, then she was dead." The family had a funeral "and I was able to grieve her death once, and it doesn't come up again and again like Leonard's." Without having a body for a funeral, "the finality was never there," she told me. "I mean it still is not there. His remains have never been found. So you never quite knew are you grieving because he worked there? Are you grieving because he is dead? At some point you do come to realize that he is gone. He is dead." But, she explained, grief and mourning are not simply about the death of a loved one, they are also about the uncertainty that accompanies it.[101]

Castrianno, who later moved to Omaha, Nebraska, has dealt with complicated grief by becoming a public voice and advocate for 9/11 victims and families in her state. She realized that because her fellow Nebraskans were so far from New York City, once the initial news coverage of the attacks subsided, they largely went back to their own lives and only thought about 9/11 near the anniversary of the attacks. Castrianno feels compelled to remind people of the people who lost their lives on 9/11 and her belief that terrorism is not something that just happens in New York or Washington—it can happen anywhere, even in Nebraska. As such, she organizes a yearly memorial in Omaha in which a flag is planted for each victim in a field on the grounds of Memorial Park—which is dedicated to individuals from Omaha and surrounding communities who have been killed or went missing in the line of military duty. She also speaks about the World Trade Center attacks at gatherings around Nebraska.

Trajectory of the Identification Effort

As part of their ongoing communication with families throughout 2002, by the early summer, OCME personnel were beginning to warn relatives that while the number of identifications had increased dramatically in early 2002, they would fall off just as rapidly later in 2002. At this point, the number of remains with usable STR profiles would dwindle and the cases remaining would be far more complex due to missing or erroneous antemortem information, incorrect kinship data, and other issues. By September 10, 2002, the OCME announced that it had identified 1,401 people—roughly half of the victims of the World Trade Center attacks. Chief Medical Examiner Dr. Charles Hirsch told the press and families that their ultimate goal was to identify 2,000 victims, but he vowed not to stop if they got to that mark. By June 2003, 1,508 people had been identified. In the next year and a half, only seventy-seven new identifications would be made, mostly via DNA alone. That said, during that period, nearly 3,000 remains were associated with a previously identified victim.[102]

Throughout 2002 and into 2003, the OCME continued to reassure families that identifying and returning the remains of as many victims as possible was its ultimate goal—they did not envision resting until all available technologies and techniques had been exhausted. On the first anniversary of the attacks, Hirsch promised families, "After we exhaust the limits of current science, if something new or better comes along, we'll do that. We will never give up."[103] In their January 25, 2003 update, for instance, the OCME reassured families that much-discussed local budgetary woes would not impact their work. "Our WTC operations are funded through FEMA and are not impacted by government cutbacks in the city or state of New York. We know that our process will continue for at least another year and in all likelihood, much longer."[104]

The OCME also highlighted the new techniques that it was using on the World Trade Center remains, particularly those that allowed for the creation of DNA profiles from smaller fragments of DNA. The first was Bode-Plex, a modification of STR testing that was being developed specifically for the World Trade Center by Bode Technology, based on work done at the National Institute of Standards and Technology (NIST). The second was single nucleotide polymorphisms, or SNPs, which examine minute variations in DNA structure as opposed to the number of repetitive STR elements in a given region of the genome. While neither of these techniques had yet been validated for forensic use, the OCME hoped that they would soon. The OCME also noted that it was working on the creation of virtual profiles, which involved the combination of several incomplete profiles based on overlapping markers. The basic idea was that statistical certainty of a match could not be reached from the incomplete profiles, but it could if they were combined.[105]

Yet, as the effort dragged on, it was becoming increasingly clear that each innovation brought only a few new identifications. Although these events were clear victories for the families involved and the OCME, they were not making a huge dent in the number of remains being identified. The number of remains that would benefit from these new technologies was, quite simply, small.

On April 22, 2005, Hirsch wrote a letter to families formally announcing the "pause" in the OCME's identification efforts that had been discussed with them over the past several months.[106] In a speech at City Hall a few weeks later, he noted that in three and half years,

> we have identified 1,591 of the 2,749 victims, and have identified 54% of the 19,195 recovered remains. Of the identifications by a single

> modality, 86% have been by DNA, and DNA has been a component in 89% of the identifications made by more than one modality. Without modern DNA technology, we would have identified only 741 WTC victims; 844 families that now have identification would have none, and we would have no hope of making additional identifications. Through the efforts of our personnel and private companies, we have advanced DNA technology to its present limits, and we have virtually exhausted those limits. Consequently, we now have reached the point where additional identifications are unlikely with current technology.[107]

In describing the situation as a pause, Hirsch made it clear that the OCME was not abandoning the effort entirely, but was rather waiting for the development of new methods that would allow viable analysis of DNA from badly damaged human remains. Family reactions ranged from appreciative of the effort that the OCME took, to sadness and resignation that more than 1,100 victims could not likely be identified baring the development of some amazing new technology, or the discovery of some large cache of remains that had previously been overlooked in the recovery effort.[108]

True to their word, the OCME did ultimately restart the identification process in September 2006 when Bode Technology introduced a new method for extracting DNA from damaged bone samples. Informing families of this development, Hirsch concluded, "my colleagues and I reiterate our commitment to you: we will never quit."[109] As of the June 2007, which appears to be the final formal announcement by the OCME on this issue (despite promises of monthly updates), 2,305 samples had been sent to Bode, of which 1,919 had been subjected to the new extraction procedure. Of those, more than 50 percent yielded a usable DNA profile. Most importantly, this new technique led to twelve new identifications. Additionally, 865 remains matched previously identified victims, 5 remains matched previously identified terrorists, and 67 matched previously recovered bone for which there was no link to a known victim. There were also sixteen new profiles not linked to any other previously recovered remains or known victims.[110] In another project, the OCME contracted Cybergenetics, a company that marries forensic science and data analytics, to go back through all available DNA profiles to look for matches that human analysts might have missed. The company provided the OCME with information about matches it found so they could be further investigated.

Long-Term Storage

Having declared early on that it would not stop the identification efforts until all remains that could potentially be identified were identified, the OCME was responsible for storing these remains and periodically retesting them as new technologies became available. Further, for the families that declared that they did not want any additional remains returned to them, the OCME had to store those remains in perpetuity, or until such time that the next of kin changed their mind. Many families received small fragments of their loved ones' bodies early on and declared that they did not want to be notified about additional remains. In several cases, much larger fragments (and in one case a nearly complete body) were recovered, but the OCME was legally bound not to say anything to families. These bodies and body parts thus became the permanent responsibility of the OCME.[111] There were also many families who simply did not want to receive any more remains after the initial finds.

In order to preserve the remains for long-term storage in a way that did not violate the rules or values of any religion or culture—several religions oppose the chemical preservation or treatment of bodies—the OCME decided in 2003 to dry them out using purpose-built heating rooms. In this way, they could be stored in a sanitary and respectful manner at room temperature either until they were identified and claimed or placed in perpetual care of the 9/11 World Trade Center memorial that was yet to be planned.

The OCME hired Kenyon International Emergency Services—a private contractor with extensive experience in responding to disasters—to conduct this drying process on everything from small pieces of bone and tissue to nearly complete bodies. The desiccated remains were then placed in special plastic shrink-wrap with identification tags both on the inside and the outside of the bag. If there was no identification yet, the remain's case number was on the tag. If the remains had been identified, then the victim's name and missing persons case number was placed on the tag.[112] These identified or unclaimed remains were housed at Memorial Park until late October 2013, when Hurricane Sandy caused them to be removed to an undisclosed location. The remains now rest seven stories below ground, at bedrock of the World Trade Center site in OCME's repository, which is located behind a large wall of the National 9/11 Museum in between the footprints of the two towers.

While the massive identification effort certainly demonstrated to families that the OCME cared about them, it is an open question whether the extreme measures taken to analyze miniscule and highly damaged remains were worth it. Have families benefited from continual identifications? Is it really therapeutic to receive a bone fragment the size of a thumb or even a thumbnail? Might it not have been better to bury the remains that could not be easily identified together in a respectful common grave? Was the emotional toll taken on so many OCME employees who worked on the World Trade Center worth it? These are questions that do not have a single answer. Ribowsky argues that the "staggering" $80 million dollar price tag for the effort was "well worth it" for several reasons: first, because of the relief it brought to families whose loved ones had been identified; second, the care provided to the dead illustrates the strong moral and ethical compass of the American nation; and finally, because it led to tremendous advances in DNA identification technology and "changed the way governments around the world respond to mass fatalities." Yet he notes that the World Trade Center effort may have been a "Pandora's Box." Given the lengths that the OCME went to, he notes, "victims' families now demand that more be done to locate and identify their loved ones" in the wake of conflict and disaster.[113] Unfortunately, not all medical examiners are equipped for the task and the federal government is unlikely to pick up the tab for all but the most dramatic and nationally significant events. Moreover, the identification of the dead is only a small step in the process of individual healing and social reconstruction. How the dead are remembered matters just as much as the way their remains are handled.

4

Master Plan

On the morning of September 11, 2001, Monica Iken was looking forward to spending the next day with her husband, Michael. It was to have been their second anniversary. Like usual, Michael had left home early in order to commute from their home in Riverdale, New York, to the World Trade Center, where he worked as a bond trader for Euro Brokers Investments on the 84th floor of the South Tower. On September 12, Iken found herself on an empty Metro North train heading into the city armed with photographs that she hoped would help her to locate him. When she emerged, bleary-eyed and grief-stricken, from the Grand Central Terminal onto 42nd Street in Manhattan a photographer from *Newsday* snapped her picture. It appeared on the front page of the September 13 edition of the paper. Instantly, she became a public face of the true terror of 9/11. Iken had gone into Manhattan to find out if anyone had seen Michael in the hospital—there was a possibility that he had been rescued after being struck on the head with debris that rendered him unconscious or unable to remember who he was. The chances were slim, Monica knew, but she felt she had to do whatever she could to find him.[1]

Monica had spent the previous twenty-four hours in a state of shock and disbelief. She was now struggling to find the person she loved most in life. Michael had called her shortly after the first plane struck the North Tower to let her know that he was okay, and to tell her not to worry because he was not in the tower that was affected by the crash. They then spoke again a few minutes later and had a similar conversation. Not quite comprehending what he was telling her, she switched on the television just in time to see

United Flight 175 crash into the South Tower. When she tried to call him back to see if he was okay, the line was dead. She spent the entire day waiting by the phone. Each call that was not him brought further anguish and disappointment. She would never speak to Michael again. To this day, his remains have not been recovered.[2]

Once it became apparent that Michael was no longer alive, Monica got into her bed and stayed there for several days—barely sleeping and not eating. As she tells the story, one day around two weeks after the attacks, she had a horrible vision that the Port Authority and New York developers were going to rebuild the World Trade Center complex on top of Michael's body and the remains of the thousands of other people who had lost their lives in the attacks. Barely two weeks after the attacks, people were already talking about how the site was going to be rebuilt. At that point, she got out of bed and declared to her family and friends that she had found a mission. She would prevent the site from being rebuilt as if nothing had happened, where people would work and shop and life would return to normal. Rather, she would ensure that a memorial was built to honor those who lost their lives there. They told her she was crazy—that she was still grieving and not thinking clearly, that she was only one person and couldn't stop the redevelopment of the site by some of the most powerful people and organizations in the world, that there was no way that some of the most prime real estate on Earth would be left fallow and economically unproductive. But she was undeterred. Somehow, she believed, she would stop them from building "over dead people." She told me that, in the early days after 9/11, she would have chained herself to the fence around the site to stop construction if she felt that that was the only way to stop it. In her view, the entire sixteen acres of the trade center site had to be preserved as a memorial befitting those who lost their lives there on September 11—to do anything else would be tantamount to defiling the graves of the dead.[3]

Soon after her vision about the rebuilding the World Trade Center site, she founded the organization September's Mission, and began figuring out what it would actually mean to protect the sixteen-acre World Trade Center site from development. She had ample media recognition from being on the cover of *Newsday*, but she had no clue how to run a public relations and advocacy campaign. Her sister, who worked for the *Washington Post*, invited her to come to Washington to get advice from her friends and associates on how to start the organization and to learn about the process of establishing public memorials. Over the next several weeks, September's Mission began to take shape.[4]

Differing Visions

By the end of September, it was already clear that the effort to redevelop the site would not simply be about money and power, although there was plenty of that to go around. It would also be an ideological battle. On September 30, the *New York Times* printed an article that laid out what a few eminent artists and architects thought should be done at Ground Zero.[5] Their positions fell into three distinct categories: (1) don't rebuild at all and instead make the site a park-like memorial (either a soothing one with trees and fountains or a barren one that constantly reminds people of the absence of the towers and the people who died when they collapsed); (2) turn the site into a mixed-use venue that has office space, cultural venues, and a suitable memorial to the victims of the attacks; and (3) rebuild bigger and better as a symbol of defiance, with only minimal recognition of the tragedy that took place at the site.

The park version of the first category was best represented by artist Shirin Neshat, who said, "It would be absolutely cruel to build a building on the site. In order to remember the loss of lives, you need a certain amount of emptiness. If you build, it's like you're covering up the tragedy and will forget it. Visually, what I see is something very spare and meditative, an open space like a park, but with the names of every single victim written on the ground. The ground would be stone, a very big circle or square, and flowers would surround it." For sculptor Joel Shapiro, who designed artwork for the Holocaust Museum in Washington, "leaving the space empty would be the most effective remembrance. It's like Berlin. You see the devastation. There are areas of Berlin that have not been rebuilt, and that's much more potent than any stupid monument you could build, because you have this real sense of what happened. We don't need a monument. You see a monument and you don't think of anything."[6]

Representing the second perspective, sculptor Louise Bourgeois stated, "This is a valuable a piece of land and it's idiotic to say turn it into a park. A park doesn't pay for itself. The land is too valuable to be left unused. That we will have to make a memorial is obvious. The people who suffered at the hands of this catastrophe must be remembered by name. The memorial should be a list of the victims' names; names beautifully hand-carved into stone."[7]

Representing the third view, sculptor James Turrell stated, "I feel we should rebuild. I am interested in seeing the working culture of New York continue. People want a memorial now because they're feeling emotional, but emotion passes, all emotion passes, and then the memorial has no meaning. The new buildings should be higher than the old ones, and there should be three of them. We should not feel bad about building on top of the ashes. All cultures are built on top of earlier cultures." Similarly, the architect Richard Meier said, "The site should not be a park. We already have a great new park along the west side. A park is not an appropriate symbol of what happened here. We need office space, we need new buildings that are an even greater symbol of New York than what was there before." For Robert M. Stern, the symbolism of rebuilding stretched far beyond New York. The skyscraper represents what it means to be American, and patriotism almost demands that we rebuild the site bigger and better than it was before. In his view, it would not be appropriate to leave the site barren or to memorialize the towers as two voids in the ground.[8]

For Iken, the notion that any building would take place on the World Trade Center site was too much to bear. Many individuals, families of the dead and missing, and family organizations agreed with her. For them, the site was akin to a cemetery and it was nothing less than sacrilegious to build atop such a space. She gained a surprising ally in Mayor Giuliani, who at his farewell address on December 27, 2001 publically changed his position on the redevelopment of the site, declaring that economic development was the wrong way to think about what should come next at Ground Zero. Instead of the mixed-used development that he advocated for in the first few months after the attacks, he stated that the only thing that should be built on the site was a "soaring, beautiful memorial" that paid tribute to the victims of the attacks, those who survived, and the rescue workers who risked their lives to save people at the scene. If the memorial was done right, he argued, the office space, the tourism, and the residential amenities would simply fall into place.[9]

Larry Silverstein, the local developer and real estate magnate who had recently negotiated a ninety-nine-year lease on the Trade Centers with the Port Authority, disagreed. Very soon after September 11, he announced that he would rebuild the Trade Center towers in some form and he launched a public relations blitz to convince the public that this was not only his right, but his duty.[10] Silverstein claimed that he wanted to heal the city's wounds and deny the terrorists a victory while he rebuilt what was supposed to have been the crown jewel of his real estate collection. His actions and words, however, were seen as greedy and insensitive to the families of victims and those who had survived the attacks at the site. Architecture critic Paul Goldberger notes that, while Giuliani was invoking the language of sacred ground, Silverstein was invoking the language of money and real estate.[11] Indeed, within a few days of the attacks, Silverstein had devised a scheme to maximize his insurance claim by declaring that the attacks were two discrete events, not one. If accepted by insurers, this ploy would entitle him to collect double the payout—around $7 billion, rather than $3.5 billion. He had also contacted his architect, David Childs, within days of the attacks to request that he devise a plan to rebuild the World Trade Center as soon as possible so that businesses could return (and begin paying rent again) quickly.[12] His concerns about money were not a surprise. Silverstein was still required to pay the PA $120 million a year in rent even though the towers had just been destroyed. Further, the Port Authority argued that Silverstein was obligated to return over 10 million square feet of office space to the site, along with the 600,000 square feet of retail space recently leased to the Westfield Group for a ninety-nine-year term.[13]

The Lower Manhattan Development Corporation

Silverstein had an ally even more powerful than Giuliani: Governor George Pataki. All the federal funding dedicated to the cleanup and redevelopment of the World Trade Center site flowed through the governor's office. On November 2, 2001, Governor Pataki announced the creation of the Lower Manhattan Development Corporation (LMDC), a new public entity within the state-run Empire State Development Corporation (ESDC) that would take the lead on planning and reconstructing the World Trade Center site and surrounding areas, and would serve as a repository for the billions of dollars of aid coming from the government.[14]

Making the LMDC a division of ESDC was a shrewd political move by Pataki. ESDC is a state agency and public benefit corporation set up to advance economic development and job creation within New York State. Its main function is to promote the creation of new businesses and economic activities (e.g., film production, winemaking, manufacturing, etc.) throughout the state. The leaders of the ESDC are appointed by the governor and are generally his political allies. The ESDC, at least in theory, operates independently of other governmental structures in accord with its pubic mission. This means that the agency enjoys a tremendous amount of power without requiring the approval of the state legislature or even the consent of local officials. This feature of the ESDC was especially important in the context of 9/11 for two reasons: first, it was widely believed in the weeks after the event that Democrat Mark Green would win the upcoming mayoral race to replace Giuliani; and, second, the state assembly was controlled by Democrats and their leader Sheldon Silver, who represented the district that included the World Trade Center. Republicans did not want a Democratic mayoral administration or state assembly to have too much control over the redevelopment effort. Pataki and his allies wanted to promote business interests in order to spur the city's economy. They believed that doing anything else would be disastrous for the New York City region and the state. In a constrained economy, businesses would either wither or move their operations to New Jersey or Connecticut.[15]

Pataki's political motivations become even clearer when one realizes that the initial LMDC plan gave the governor of New York control of six out of nine seats on the agency's board of directors, with the city controlling the other three. Shortly after Republican Michael Bloomberg pulled off his surprising victory against Green, however, the board was quietly expanded to sixteen, with the city and the state each receiving eight slots.[16]

Most appointees had backgrounds in finance or business, and many were or had been affiliated with the Pataki administration or New York City government.[17] Typifying this trend, the LMDC board was chaired by Pataki appointee John Whitehead, who had, at various points of his long career, served as cochairman of Goldman Sachs, director of the New York Stock Exchange, deputy secretary of state under Ronald Reagan, chair of the Federal Reserve Bank, and chairman of the Mellon Foundation and the Harvard Board of Overseers. He had also served on the boards of numerous philanthropic, arts, and educational organizations. Among the most notable LMDC board members were the former chairman and CEO of NASDAQ (Frank Zarb), the current chairman and CEO of the New York Stock Exchange (Dick Grasso), and Roland Betts, owner of Chelsea Piers and former co-owner of the Texas Rangers (with his close friend George W. Bush). Giuliani's appointments were overwhelmingly current or former city officials.[18]

Early on in the process, it was unclear what role the LMDC would play. It was charged with redeveloping the World Trade Center site and providing assistance (derived largely from federal money) to the businesses and residents of lower Manhattan as they struggled to rebuild lives in what was tantamount to a war zone. What was clear was that many people wanted a say in the rebuilding efforts, and the redevelopment effort would be fraught with political, emotional, and economic tension. Monica Iken was initially joined by many individuals and groups, including the 9-11 Widows and Victims' Families Association (which was run by Marian Fontana and was made up primarily of uniformed service widows), in calling for a slow, methodical effort to recover the remains of the dead and no future building on the site. But not everyone believed that moving slowly was a reasonable option—or that the site should be treated as a cemetery.

Prominent art gallery owner Max Protetch began asking architects from around the world to propose ideas for the redevelopment of the World Trade Center site. Protetch had a strong reputation in the contemporary art and architecture world and regularly showed art from the most important American and Chinese artists at his gallery in New York. His goal in organizing the exhibition was to provide a forum for designers to express a range of ideas of what might eventually be built on the site—as a way of encouraging both the healing of the city and artistic excellence.[19] An exhibit of these proposals ran at his gallery, which also specialized in exhibiting and selling architecturally oriented drawings and objects, from mid-January through mid-February 2002, and then at the National Building Museum in Washington, DC from April through June 2002.[20] This effort was undertaken entirely independently of the LMDC, the families, or any government agency, cultural organization, or professional association.

Many local planners and architects were upset that Protetch did not consult them, and many families found it deeply troubling that all the proposals started from the assumption that new buildings would go up at the site. And, if buildings were to be erected at the World Trade Center site, what kinds? Early on, Sally Regenhard, whose twenty-six-year-old son, firefighter Christian Regenhard, was killed as the towers collapsed, formed the Skyscraper Safety Campaign, arguing both for better safety standards in tall buildings, and against the return of skyscrapers to the site. At the same time, other groups emerged as forceful advocates for rebuilding the twin towers much as they had looked before the attacks—or taller.

The Community Board #1 Town Hall Meeting

Just how stark the differences of opinion were became clear at the January 29, 2002 town hall meeting held at Stuyvesant High School by

Manhattan Community Board #1 (CB 1), a city-chartered organization designed to give residents and businesses of lower Manhattan (defined as the Financial District, Battery Park City, the Civic Center, Tribeca, Greenwich Village South, and the South Street Seaport) an opportunity to have a say in any major planning initiatives being undertaking in the area.[21] This meeting made plain the disagreements that existed between residents of lower Manhattan and relatives of the victims of the 9/11 attacks. It also revealed forcefully that not all relatives felt the same way about the redevelopment of the site. As CB 1 Board Chair Madeleine Wils (also an LMDC board member) noted in a 2002 interview for the Public Broadcasting Service documentary *America Rebuilds: A Year at Ground Zero,* lower Manhattan was the fastest-growing sector in New York City. This vitality was lost the day the towers fell and residents wanted it back. Residents spoke passionately about wanting to get back to a sense of normality, both for themselves and their families. Children wanted to know when the site would be cleared and when they could once again shop at Borders Book Store. Their parents wanted the amenities that they had grown used to, such as farmers' markets and coffee shops, to reopen.[22]

Iken, on the other hand, delivered a forceful speech in which she implored the powers that be not to build on top of the place where her husband died—how can you worry about the reopening of a book store or a fruit stand, she asked neighborhood residents, when "I lost my husband, I lost my future, I won't ever have children?"[23] In a subsequent interview, she said that she found the town hall meeting particularly upsetting because residents "were so concerned about things that I would have never thought I would hear from people. They had no remorse, no compassion for the families that had to watch their loved ones be murdered. And what that means for the families that are left, the survivors and the rescue workers who have to deal with that. . . . It was very discouraging to actually attend a town hall meeting of that nature and see how people just didn't understand what we were going through."[24]

Not all relatives went as far as Iken in demanding that all sixteen acres be preserved. Lee Ielpi, who lost his son, firefighter Jonathan Ielpi, and was a key player in the effort to recover human remains at the site, stated that he had come to the community forum with the idea that the entire sixteen acres of the World Trade Center should be preserved as a memorial to those who perished in the attacks. After listening to the residents of the area talk about their own pain and suffering, he said he began to realize that his idea was "grandiose," and that the community had as much of a right to rebuild as he did to have a memorial to his son and the other victims of the attack. In the end, he told the community that he respected their right to move on and recognized that this meant that the site ought to be developed.[25] Ielpi would soon emerge as a proponent of mixed-use development, with the most meaningful parts of the site (especially the footprints of the towers and the Marriot) being held sacred and reserved for a memorial, with tasteful development taking place on the periphery of the site. In time, Monica Iken's position would moderate too, but just a few months after losing her husband Michael, the pain and desolation was still too intense for her to fathom new buildings emerging from the site where he was murdered.

Listening to the City

On February 7, 2002, the recently created Civic Alliance to Rebuild Downtown New York attracted more than 600 participants to its first event, called "Listening to the City," at the South Street Seaport. The Civic Alliance, as it was generally called, was a partnership between the Regional Plan Association (a group of New York architects and planners) and The New School, New York University, and the Pratt Institute. It was made up of more than eighty-five business, community, design, and environmental groups that wanted a say in how lower Manhattan was redeveloped in the aftermath of 9/11. The goal of Listening to the City was to solicit feedback from a broad swath of area residents and workers and those affected by the World Trade Center attacks on how to chart a "a bold new vision" for lower Manhattan while still honoring those who died in the attacks.[26]

Based on self-reported demographic information, the participants did indeed come from diverse backgrounds that, while not quite matching the area's demographics in all domains, still captured the diversity of the neighborhoods surrounding the site (particularly in socioeconomic background). Further, about 5 percent of the participants were family members of victims, 5 percent were rescue workers who responded to the attacks, more than 15 percent were survivors of the attack, and nearly 25 percent were at or near the World Trade Center on 9/11. The Civic Alliance noted that in subsequent activities, they would try to do a better job of recruiting a higher proportion of black and Hispanic participants, as well as people who lived outside of Manhattan in the other four boroughs, New Jersey, Connecticut, and the rest of the state.[27]

After the day-long event, which included facilitated small group discussions and computerized technology that allow viewpoints to be collected and tabulated in real time, the Civic Alliance confidently proclaimed that the participants "had forged a common vision of the values and principles for rebuilding that represent the aspirations, memories and pride of New York and the metropolitan region."[28] That common vision called for balance among: (1) filling social, cultural, and residential needs and the push to restore real estate; (2) strengthening the financial sector and broadening the economic base of the neighborhood; and (3) moderating the desire to rebuild as quickly as possible and undertaking a careful, deliberative planning process.

In general, there was widespread agreement that lower Manhattan should become a "vibrant, 24-hour, mixed-use community"; provide better transportation links to the rest of the city and the region; offer equal opportunities to business ranging from investment banks to shoe repair shops; increase its stock of affordable housing; encourage the growth of cultural institutions; provide enhanced residential amenities like libraries and community centers; and include more green space, particularly along the waterfront.[29] There was no consensus on how exactly the World Trade Center should be rebuilt. It was clear, though, that most people involved in this town hall forum wanted their voices, and the voices of fellow citizens, to be heard in the process.

In addition to a shared vision for lower Manhattan, the Civic Alliance proclaimed that participants "expressed a common vision for a powerful memorial that is integrated into the very fabric of downtown. This memorial would honor the 'everyday people' who were lost, as well as the heroism, sacrifice and resiliency that were—and continue to be—demonstrated throughout the city, region, nation and world."[30] A careful reading of the views of participants suggests, however, that the consensus that a memorial be built did not translate to strong agreement over what the memorial should look like, or who it should honor. For some participants, the memorial ought to be primarily for those who died. For others, it should soothe the emotions of those who survived and relatives of the victims of the attacks. For still others, the "heroes" of 9/11 deserved special recognition as well, and for many participants, the memorial ought to be for anyone who was adversely affected by the attacks, whether they were in the area when the planes hit or saw the attacks on television from thousands of miles away. Further, there was a sense that the memorial ought to honor the "American values and ideals that were attacked—and the innocence that was lost—at 8:46 am on that bright September morning." Some went so far as to suggest that the Trade Center itself, and "what it stood for as a symbol of our global village" ought to be memorialized as well. Some commonly expressed meanings included liberty, community, and democracy.[31]

Views about what tenor the memorial ought to strike were equally diverse. Some participants wanted to focus on the individuals who lost their lives; others wanted to celebrate the courage, sacrifice, heroism, resiliency, altruism, volunteerism, and duty shown in the response to the attacks; and still others believed that the true horror and shock of the event ought to be captured for future generations. As one discussion facilitator, who also lived downtown, noted, questions about memorialization of the event and its victims "plunged us back into grief, and the answers were unsatisfyingly vague. We were reminded how difficult these questions are for all New Yorkers, and how much more we will have to share to find the answers."[32]

When asked what the memorial should look like, participants offered a wide range of ideas and sketches, some of which were conventional in nature and others of which were "quite innovative and new."[33] There was widespread support for explicit remembrance of the victims of the attacks, including "names, characters and biographies" of all who were killed in the attacks. After that note of unity, a wide array of proposals was presented. Some argued that there ought to be natural elements present (i.e., trees, water, grass, etc.), and others that the memorial should include light and illumination. Many wanted quiet spaces for contemplation and prayer, and others believed that there should be an educational component to the memorial. Some participants also advocated for private spaces for families of the victims to be able to visit as well as the preservation of part of the façade of the towers. Many participants also noted that they believed that the memorial ought to be "firmly situated on the footprint of the two towers."[34]

Ultimately, almost everybody who participated in the event hoped that decision makers and planners would listen to the feedback they offered on the redevelopment process. The Civic Alliance claimed to be working closely with the LMDC, and Louis Tomson, LMDC's president and executive director, claimed that the agency would be paying attention to the Alliance's reports from the first Listening to the City event and the second, much larger, one that was planned for July, and it that it would take its findings into account during the development process.

Getting Started at the LMDC

Conversations with many stakeholders and the early feedback coming out of the Listening to the City event convinced LMDC board chairman John Whitehead that the corporation would have to strike a balance between rebuilding and restoring commerce on the one hand, and constructing a suitable memorial to honor the victims of the attacks on the other. In order to solicit public opinion, and to make it seem like the LMDC was not dictating the redevelopment of the site, the corporation organized advisory councils to help guide the decision-making process. These centered around stakeholder groups such as arts, education, and tourism; professional and financial services businesses; residents; restaurants, retailers, and small businesses; commuters and transportation planners; and, perhaps most importantly, families of victims of the attacks. This decision was cloaked in practical and patriotic terms by the LMDC. In a July 16, 2002 speech, LMDC president Louis Tomson explained why the agency was going to such great lengths to gather public opinion: "While it is too early to say what will one day rise from the World Trade Center site, we are certain of one thing: what we create together will be a testament to—and consistent with—the very principle that came under attack on September 11th: Democracy."[35]

While most of these advisory committees would meet only a few times, the Families Advisory Council (FAC) met at least thirty-five times from February 28, 2002 until it was disbanded in early 2006.[36] Even at the first meeting it was clear to the families of victims that their views were only one of many perspectives that the LMDC would address. In early meetings, family members involved in the FAC were making demands that did not necessarily fit into the LMDC's long-term plans. Most notably, a vocal subset wanted to prevent any development on the sixteen-acre site. Other families and family groups sought to force the Port Authority to follow New York City building codes if new buildings were put up on the site (because the PA was an interstate agency, it was exempt from the city's building codes); preserve the footprints of the towers down to bedrock without the intrusion of utilities or transportation infrastructure; and prevent the reconnection of any of the main roads that had been blocked off when the World Trade Center was originally constructed.

Within a few short weeks of the creation of the Family Advisory Council, many relatives of victims began to feel that the LMDC was soliciting input from them without intending to actually listen to it. It was clear that cynicism and mistrust were already building between many relatives and the LMDC. Monica Iken believed that the meetings were not a genuine conversation since LMDC representatives generally told families what was happening but did not offer them a chance to influence the redevelopment effort. For instance, families were not asked to give feedback on the plan to continue running PATH and subway trains through the site even though it was, in Iken's words "our graveyard." Echoing the sentiments of many families, she argued that the LMDC was moving far too quickly and was not considering alternatives to the plan to replace all of the retail, office, and hotel space in the area. In her view, "the public is being misled to believe that families are having input and now the families are realizing that they haven't had any input on any of these decisions and they're going at a speed that's very fast."[37]

Family Associations Unite

Many of the family groups that were taking a strong interest in the recovery efforts at the World Trade Center mobilized around issues related to the redevelopment of the site and the memorialization of victims. In order to make sure their voices were heard, several groups and individuals came together in early 2002 to form the Coalition of 9/11 Families. The initial coalition was comprised of the 9/11 Widows and Victims' Families Association (Marian Fontana and Lee Ielpi), Give Your Voice (Jenny Farrell, Michael Cartier, John Cartier, and Bill Doyle), September's Mission (Monika Iken), the Skyscraper Safety Committee (Sally Regenhard), St. Clare's WTC Outreach Committee (Dennis McKeon and Patricia Reilly), Voices of September 11 (Mary Fetchet and Beverly Eckert), and the WTC United Family Group (Anthony Gardner). This group would eventually expand to include most of the families that engaged publicly about the recovery, identification, and memorialization of the victims of 9/11.

Although each organization had its own central mission, they shared a common belief that families ought to have a significant voice in determining the fate of the World Trade Center site, that the primary objective of any development of the site ought to be to honor the lives lost on September 11, and that the entire sixteen-acre site ought to be treated as a "hallowed burial ground," even if buildings were ultimately built on part of the site.[38] In their first newsletter, they laid out their recommendations, most notably that:

- "as much acreage as possible, including the WTC Tower 1, Tower 2 and the Marriot must be entirely devoted to a Memorial and should have permanently protected legal status. This includes the area inside the slurry wall quadrant of Greenwich Street (East side), West Street (West side), Vesey (North side), and Liberty (South side)."[39] According to the coalition's calculations, this area totaled nine-and-a-half acres and represented the space where the most human remains were recovered according to the FDNY's GPS mapping effort. Thus, the coalition associated the location of remains with the sanctity of the space and its importance to the memorial.[40]
- "The overriding purpose of architecture and landscaping at the site should be to make visitors aware of the enormity and reality of the events on 9/11 and the personal sacrifices of those involved."[41]
- "Treatment of the site should be parallel to sites of comparable importance in American history, such as Pearl Harbor, Gettysburg and Normandy that contribute to the economic, cultural and social values of New Yorkers in the manner of Central Park, Lincoln Park and the Mall in DC."[42]

Additionally, coalition members advocated for the long-term storage of unidentified and unclaimed remains at the World Trade Center site. This facility ought to accommodate continued identification efforts by the OCME, and provide a respectful and dignified resting place for the remains as well as a quiet, private area for mourning and reflection by families who lost loved ones in the attacks.

Overall, the coalition argued that consideration for the victims and their families ought to drive any redevelopment of the site. Such recommendations were clearly not in line with the desires of residents and businesses, most of whom preferred not to be constantly reminded of the deaths of so many people in the space where they were living and working. While they certainly wanted a dignified memorial to the dead, they wanted it to be subtle—just one aspect of the redeveloped site. Residents stated over and over again in public forums that they did not want to live with a cemetery or a "giant mausoleum" in their backyard.[43] As local resident Jeff Galloway stated in the July 2002 Listening to the City II forum, "I believe a memorial is appropriate, but I don't want to live in a cemetery, where you have no choice but to cry. I want something triumphant, showing how we took a direct hit and survived, that they didn't beat us. The people that I knew who died, they wouldn't want us to mope."[44] Over time, however, the families and residents would come to find that they could agree on many more things than they originally imagined—particularly with respect to the extent to which the LMDC was willing to seriously consider their views on the redevelopment of the site.

Of particular concern to the coalition families was that the LMDC seemed to be making decisions without legitimately consulting the Family Advisory Council. In the families' view, representatives from the Port Authority would come to FAC meetings to tell members about the restoration of PATH service between New York and New Jersey, and LMDC representatives would describe the proposed reopening of Greenwich and Fulton Streets without asking for meaningful feedback from the families, many of whom had strong feelings about these plans.[45]

According to families, this lack of meaningful consultation impeded constructive discussion about the future of the site—and would lead to disagreements within the FAC between family members who opposed such plans and those who did not. This contentious dynamic, and the feeling that certain family members created trouble, frustrated LMDC and local officials, and efforts were undertaken to diffuse the tension.

One such effort was the decision to invite Edie Lutnick to join the FAC. Lutnick, a lawyer with an MBA, lost her brother Gary in the attacks. Gary was a bond trader at Cantor Fitzgerald, the financial services firm located on the top floors of the North Tower that lost 658 employees on September 11—far more people than any other organization. Edie had another family connection to Cantor Fitzgerald: her other brother Howard was its CEO. At Howard's request, Lutnick disbanded her solo law practice and set up a relief fund for affected Cantor families, to be funded primarily by the firm's profits. In this role, she not only distributed money to families in need, but also became a public advocate for them.[46]

According to Lutnick, the LMDC brought her in to counteract what they saw as the coalition's unreasonable demands. Looking back, she believes she was supposed to side with prodevelopment interests, allowing the LMDC to state it had family support. But Lutnick found herself agreeing, or at least sympathizing with the demands of the coalition families:

> While I can understand that there would be cost considerations to doing away with the Port Authority building exemptions, and that 16 acres of prime real estate dedicated solely to a memorial is prob-

ably unlikely, preserving the land where 2,479 people were murdered and treating it as sacred ground, down to bedrock, which is where a majority of the human remains were found and are the historic ruins of the building, not allowing cars to drive across what is in essence a massive cemetery, and ensuring that no matter what goes on top it is constructed safely, doesn't strike me as patently unreasonable. I know that some of the deliveries haven't been refined, but I'm having a hard time figuring out what I'm supposed to oppose.[47]

It was at her first FAC meeting that she came to believe that the LMDC wanted the imprimatur of the families to do essentially want it wanted to and was unprepared to entertain demands that went against this fundamental plan. Put another way, the powers that be would give the families whatever they wanted so long as it did not interfere with their plans. In her view, "the problem is philosophical. The governor, mayor, and LMDC (who are appointed by the governor and the mayor) don't believe that the families have a place in the redevelopment of the World Trade Center site. The name of the agency is a telltale sign. It is the Lower Manhattan Development Corporation, and that is exactly what the WTC site is perceived as. It is a development project. New York City has received $21 billion in federal aid and $2.783 billion became the LMDC's budget to address 16 prime real estate acres in Lower Manhattan."[48]

The First Site Plan Request for Proposals

At around the time that the coalition was forming, Alexander Garvin, a noted urban planner who had been in charge of New York City's failed 2001 bid to host the 2012 Olympic Games, and had been a fixture in the city's planning community since 1970, became the LMDC's vice president for design and planning in February 2002. Garvin ruffled the Port Authority's feathers in March when he invited several architectural firms to submit proposals to serve as consultants to the LMDC. Responding firms were asked not to design buildings, but rather to offer plans on how all of the site's needs (office space, retail space, memorial, transportation, etc.) could be situated within the sixteen acres available for redevelopment.[49]

The PA was upset that Garvin thought he was in charge of the design process and forced him to rescind the request. As the owner of the World Trade Center site, the Port Authority believed it was the ultimate authority on what happened there, and that the primary goal of the tender process should be finding someone to rebuild the 10 million feet of office space and the shopping center that existed before the attacks. Apparently, the PA wanted, as much as possible, to erase the events of 9/11 from the site. For the LMDC, which was charged with acting in the public good, however, a suitable memorial and cultural activities were a priority, even if they did not meet the expectations of the most vociferous members of the FAC.[50]

After significant discussion, the request for proposal was reissued a few months later with the Port Authority and LMDC to review the proposals and select a consulting architect. Around this time, the PA and LMDC signed a memorandum of understanding in which the LMDC agreed not to try to impose a plan for the site on its own. This memorandum formalized the PA's desire to rebuild all the lost office, retail, and hotel space. The PA also stated that it would set aside part of the site for a memorial and cultural center. The memorandum made it clear that the PA would consider reopening some of the streets that were closed down when the World Trade Center was originally built, particularly Greenwich Street.[51]

When all fifteen proposals were reviewed in May 2002, the LMDC and PA chose the firm of Beyer Blinder Belle, not because of the foresight or elegance of their initial plans, but because they were aligned with an engineering firm with deep roots in transportation. The LMDC and PA then asked Beyer Blinder Belle (BBB) to generate six site plans that could serve as the basis for conversations about the redevelopment of the site (i.e., to suggest various ways that land could be apportioned for different uses). Although the LMDC and PA went to great lengths to assure everybody involved that the site plans were not akin to final designs, the public did not accept this claim and skewered the work of BBB. Mayor Bloomberg described BBB's six plans as "starting points"—hardly a ringing endorsement—when they were made public in July, after some modification and incorporation of ideas from the bids of other competitors.[52]

Indeed, all of the plans appeared to be based on the notion that some number of conventional-looking buildings with the required total of 10 million square feet of office space and 600,000 square feet of retail space ought to be grouped around a memorial area. The only major difference between the plans, other than variations on dimensions and number of buildings, was the shape of the memorial area: a six-acre rectangular memorial "park," an eight-acre version of the same design, a ten-acre memorial "square," a five-acre memorial triangle, a rectangular memorial "garden" of unlabeled size, and three smaller pedestrian-oriented plazas of unknown size.[53] The plans were widely panned by architecture critics as being, among other things, "cookie-cutter losers" with "shake and bake" urban design elements.[54]

The Coalition of 9/11 Families quickly sought to mobilize relatives of the victims to speak out against the BBB plans in public venues and at the upcoming "Listening to the City II" event in July 2002 at the Jacob Javits Convention Center. The lead story in its first newsletter noted that "as you are reading this newsletter, politicians, business leaders and special interest groups are discussing many options for the development of the WTC site with minimal input from family members. The LMDC has already published a blueprint indicating various options for the site development. The Coalition of 9/11 Families, however, believe that the location and qualities of the future memorial should largely be determined by family members with an appropriate allocation of land for the memorial and its support facilities."[55]

There were two issues of particular concern to the coalition. The first was that, in their view, the planning process was backward. Because they wanted development to first and foremost honor the dead, they felt that the memorial ought to be conceived before the site plan so that the memorial would structure the site, rather than the other way around. If all aspects of the site plan were put into place before the memorial was designed, then the memorial would be shaped by the visions of urban planners, rather than family members.[56]

The second main problem was that the call for site plans, and the plans themselves, did not respect the coalition's definition of sacred ground. As Monica Iken noted in a 2002 report to families about the development process, "the Coalition advocates the footprint as the area enclosed in both of the steel tower walls and the Marriott Hotel that was attached to World Trade Tower 1, from the bedrock seven stories down to the air rights."[57] The main proponents of this view were Lee Ielpi and Jack Lynch, who spent more than eight months in the Pit searching for their sons' remains, and who noted to other family members that the vast majority of human remains were recovered within this space, and therefore it should be considered forever sacred ground. The coalition, and Iken in particular, also expressed a strong desire for a dignified memorial that was contemplative, private and not a "Disney World" version of 9/11 in which tourists in shorts eating ice cream viewed exhibits meant to entertain rather than pay homage to the dead.[58]

While the LMDC, Port Authority, and the governor of New York were all willing to state that no commercial development would take place in the above-ground part of the footprints (Pataki told families that the footprints were "hallowed ground" and that "We will never build where the towers stood" at a June 29, 2002 meeting organized by Give Your Voice[59]), they were unwilling to make the same promise for the below-ground space because of the infrastructure needs of the site, particularly transportation.[60] They were also unwilling to heed demands of many family members to prevent vehicular traffic from crossing the sixteen-acre World Trade Center site.

It is important not to cast coalition members as uncompromising, emotionally driven zealots. Ielpi, for example, was willing to allow some of the infrastructure of the PATH train and subway lines 1 and 9 to pass through the footprints of the towers in order to demonstrate his and other family members' willingness to compromise when their views were taken into account by the LMDC and developers.[61] At the same time, it is also important to remember that as the Pit was cleared out in the months after September 11, the footprints of the towers essentially disappeared from view. The only reminders of their location were the remnants of the box columns that anchored the towers into the ground seven stories below street level. As one prominent New York architect noted: "To think of it as sacred ground is purely symbolic. And once you're in symbolic mode, you have to ask whether there are other alternatives you could consider that might be better."[62]

Listening to the City II

Pataki's promise not to build on the footprints, however, was tantamount to foreclosing on any other option since he controlled the LMDC. Further, the concerns of the coalition were amplified by Listening to the City II, in which more than 5,000 people participated over the course of sessions on July 20 and 22 and in an online dialogue that went along with the event.[63] In the final meeting report, the Civic Alliance highlighted the central finding of the meeting: that the six preliminary concepts put forward by BBB were "too dense, too dull and too commercial."[64] Further, they reflected a lack of imagination and monumentality, and reminded people of the bland government buildings in Albany, the highly planned state capital of New York State.[65]

Another significant outcome of Listening to the City II was that most participants found the memorial spaces that BBB had come up with to be poor or unacceptable. Many people stated that preserving the footprints of the towers—which was not accomplished in all of the designs—was crucial, although not all participants agreed with this sentiment.[66] Particularly in the online sessions, there was also some disagreement over whether the victims' families should play such a major role in defining and planning a memorial compared to other stakeholders such as survivors, residents, and rescue workers, but most participants accepted the right of families to have a major say in what kind of memorial was ultimately built.[67]

At the end of the day, the participants in Listening to the City II demanded boldness and rejected the PA's plan as dominated by real estate values, not symbolism. This conclusion was not surprising given that many family and friends of victims participated in the event and many other participants identified themselves as survivors of the attack.[68] The LMDC received similar feedback from the general public when it exhibited the six BBB plans in the rotunda of Federal Hall on Wall Street (a few blocks from the World Trade Center site) the next week.

The PA staunchly defended the BBB plan, arguing that they had an obligation to respect the property rights of Silverstein and Westfield. Frank Lombardi, an executive of the PA, reminded the audience that the location was not a blank slate—there were legal agreements and each side had contractual obligations to uphold. Lombardi, of course, did not mention that Pataki could have stepped in and declared the PA's demand for 10.5 million square feet of office space and 600,000 square feet of retail space unacceptable given the sanctity of the site.[69] Indeed, all of this political squabbling was useful to Pataki, who wanted to avoid coming to a final decision on the site until after the gubernatorial election in November 2002.[70]

Some participants recommended that the current lease ought to be cancelled so that planning decisions did not have to be beholden to it. As Liam Strain, a United States park ranger on Staten Island, stated in his Listening to the City II public comment, which was reprinted by the *New York Times*: "Silverstein should be bought out and the land taken from the

Port Authority and given to another public corporation. This Port Authority's stepping off point is the need to match the money-making capacity of the former World Trade Center, and I don't think that's where we should start from. We should be concerned with a memorial first, and cultural and open space for the downtown residents."[71] Such an idea was not far-fetched. In July and August 2002, PA and City officials held discussions about the feasibility of a land swap in which the PA would gain ownership of LaGuardia and JFK airports (which were owned by the city and leased to the PA) in exchange for the World Trade Center site. Given the complexity of valuing two very different sorts of property, questions about what the city would do with the World Trade Center site once it took ownership, the legal challenges associated with the lease agreements negotiated between the PA and Silverstein and Westfield, and the lack of clear political benefit for Governor Pataki in an election season, however, the deal died.[72]

Although it is difficult make definitive causal claims in this context, Listening to the City II seemed to have had an impact on the World Trade Center redevelopment effort. Shortly after the event, and the significant media attention that followed it, the LMDC and PA scrapped the BBB site plans and began to rethink the future of the site. Ultimately, at the suggestion of the LMDC's Roland Betts, the two agencies launched a request for innovative designs on August 14, 2002. As LMDC president and executive director Louis R. Tomson stated in a press release issued by the corporation, "At Listening to the City and other public forums throughout the last several months, we vowed to incorporate public input into the planning process. The invaluable public input we received is helping to shape the future of downtown. Involving additional design teams and allowing greater flexibility in the program will ensure that a variety of bold options will be introduced during this second phase."[73] The PA—preferring to rely on their own architects—hated the idea, but they had no choice after what happened with BBB.[74]

There is, of course, a different way of looking at the Listening to the City events, as well as the other public outreach efforts of the LMDC. The architecture critic Philip Nobel has argued, for instance, that such efforts were meant to keep people from protesting on the streets and to keep public discourse under control.[75] Similarly, eminent architect, urban planner, and architecture critic Michael Sorkin has argued that the heavily facilitated, LMDC-controlled Listening to the City event allowed the real power brokers to impress their agenda on the World Trade Center site while making it appear that this agenda emerged from democratic discourse. By controlling the questions asked, the LMDC determined what answers would be given. In other words, the public was reacting to prompts rather than generating new ideas. In a telling exchange at Listening to the City II between Sorkin and Carolyn Lukensmeyer, who was running the event as a representative of America Speaks (a nonprofit organization that facilitates public deliberations through town hall style events), Lukensmeyer announced that "in democracy, the people have a chance to speak." Sorkin, unable to control his anger, rose to his feet and shouted "Bullshit! Democracy means people have the power to choose."[76] Architecture critic Philip Nobel described another Listening to the City participant as saying that "this is the story of a thousand people drinking Shirley Temples and smoking candy cigarettes, and they all think they're in the back room with their Scotch and cigars."[77]

The LMDC and the PA, however, went to great lengths to convince the participants that their voices did matter, and that there was no other debate taking place behind closed doors.[78] Yet the two agencies heavily managed the public outreach campaign. To begin with, even though the PA lost the battle over control of the process, their earlier requirements still made it into the final set of desired site elements (i.e., the same program that had been roundly rejected at Listening to the City II). The LMDC and PA were looking for BBB in a prettier package.

Further, the way the process worked, the "innovative design study" was open only to licensed architects and professional planners and landscape architects, and would involve an initial request for qualifications to participate. Once received, the LMDC, the PA, and a group of advisors from New York New Visions, a "pro-bono coalition of architecture, engineering, planning and design organizations committed to honoring the victims of the September 11 tragedy by rebuilding a vital New York," would narrow the field down to ten to twenty finalists.[79] The LMDC would then determine which of these finalists would actually be asked to produce detailed site plans on September 30. The final design study would be due in mid-November and would include various drawings and models to be evaluated by the LMDC.

Other than offering a set of minimum standards for consideration, the LMDC did not state how the final five plans would be evaluated. It only noted, in legalistic language, that one or none of the finalists would be selected to continue the development process at the sole discretion of the LMDC at the conclusion of the exercise. The LMDC was also very careful to point out to respondents that "This is NOT a design competition and will not result in the selection of a final plan. It is intended to generate creative and varied concepts to help plan the future of the site."[80] It further noted that the purpose of the exercise was to generate public conversation, and not to produce a final design for the site—exactly the same statement made about the failed BBB designs. Further hedging its position, the LMDC and PA also made it clear in a press release about the request for qualifications (RFQ) to help rethink the site plan that its "planning staff and consultants, including Beyer Blinder Belle and Peterson/Littenberg, will continue to explore varied approaches to the World Trade Center site based on the new program alternatives."[81] The message was clear: the LMDC and PA were opening up to outside ideas, but would not compromise their institutional requirements.

In laying out the parameters for the request, the LMDC implied that a public and critical consensus had emerged at the Listening to the City event and other public forums. Philip Nobel put it another way. Listening to the City had digested the participants' views "into a passably official document that affirmed the preexisting assumptions of the powerful. Build offices and stores and cultural facilities. Certainly. Build a big train station? Definitely. The grid? Yes, bring it back."[82] For Nobel, the restoration of the street grid that had been closed off by the World Trade Center was homage to New Urbanism in vogue at the turn of the twenty-first century. This urban planning movement promoted walkable neighborhoods replete with public space and mixed-use development. Not coincidentally, it was also a big win for development because it would increase the amount of building frontage that could be devoted to retail and erase the old boundaries of the sixteen-acre World Trade Center site. Once this break-up was accomplished, Nobel notes, the mourning zone of the site could be limited to the old footprints of the towers since there was no longer a single special entity to preserve.

The Finalists

On September 26, 2002, the LMDC announced that six finalists had been chosen by the New York New Visions panel, out of a total of 407 individuals and teams that had answered the request for qualifications. The finalists were Studio Daniel Libeskind, Berlin, Germany (Libeskind was most famous for his work on the Jewish Museum in Berlin); Foster and Partners, London, England (best known for the design of the new Reichstag in Berlin); Richard Meier, Peter Eisenman, Charles Gwathmey, and Steven Holl (a collaboration of highly respected, internationally famous architects sometimes called the "Dream Team"); United Architects (another collaboration of well-known, if not quite as famous architects); a team led by the stalwart firm Skidmore, Owings and Merrill (whose previous work included the redesign of Penn Station in New York); and finally, the THINK team (led by acclaimed architects Rafael Viñoly and Frederic Schwartz). The LMDC also asked its longtime consultant firm Peterson/Littenberg to submit a site plan as well, bringing the total number of finalists to seven.[83]

On October 11, 2002, the finalists gathered for an intensive all-day briefing on Ground Zero, the requirements of the master plan, and what the various stakeholders hoped would come out of the innovative design study. At this meeting, Garvin offered them four hypothetical people to design for, so that they weren't exclusively focused on the needs of the PA and Silverstein: the resident of lower Manhattan who walks through the World Trade Center site on his way to Tribeca; the business traveler who arrives into the city via a train from Newark airport; the tourist who wants to see the site; and the commuter who is just passing through lower Manhattan on his way to work in Midtown. It is important to note that he did not include family members, survivors, or rescue workers among his target audience for the redevelopment effort. According to Goldberger, Daniel Libeskind was the only architect who asked to venture down into the Pit to get a feel for the site at its foundation.[84]

Studio Daniel Libeskind

On December 18, 2002, the LMDC and PA revealed the master plan proposals put together by the seven finalists at a raucous, packed event at the Winter Garden atrium in the World Financial Center across West Street from Ground Zero. The proposals varied both in terms of aesthetics and theoretical underpinnings. Leading off the presentations, Daniel Libeskind introduced his proposal, called "Memory Foundations," with a biographical note. "I arrived by ship to New York as a teenager, an immigrant, and like millions of others before me, my first sight was the Statue of Liberty and the amazing skyline of Manhattan. I have never forgotten that sight or what it stands for. This is what this project is all about."[85] He then acknowledged the differences of opinion over whether to rebuild the World Trade Center or keep the site empty as a tribute to the victims of the attacks. In the end, however, his visit to the site seemed to provide him with a way to resolve this "seemingly impossible dichotomy." For Libeskind, the appropriate metaphor for his site plan was not religious or geometric or even architectural form, as it would be for the other finalists. Instead, it was political, even patriotic. In introducing his design, he wrote:

> The great slurry walls are the most dramatic elements which survived the attack, an engineering wonder constructed on bedrock foundations and designed to hold back the Hudson River. The foundations withstood the unimaginable trauma of the destruction and stand as eloquent as the Constitution itself asserting the durability of Democracy and the value of individual life.[86]

Just as the Constitution is the key to American democracy, he argued, the bedrock foundation of the World Trade Center site is the key to its meaning, and the focal point of his design. "We have to be able to enter this hallowed, sacred ground while creating a quiet, meditative and spiritual space. We need to journey down, some 70 feet into Ground Zero, onto the bedrock foundation, a procession with deliberation into the deep indelible footprints of Tower One and Tower Two." While the foundation did indeed represent death and memory in Libeskind's plan, it also represented resilience and the continuity of life. For it was at this level that one would find not just the memorial and museum, but also transportation infrastructure that was so important to the resumption of normalcy in the city.

In addition to these below-ground spaces, Libeskind also planned two large public spaces that memorialized the victims and honored those who survived and sought to save lives on September 11: the Park of Heroes and the Wedge of Light. Finally, he planned to create a skyscraper with a towering, 1,776 feet spire that would be home to the "Gardens of the World." This tower would rise "above its predecessors, reasserting the preeminence of freedom and beauty, restoring the spiritual peak to the city, creating an icon that speaks of our vitality in the face of danger and our optimism in the aftermath of tragedy. Life victorious."[87]

Foster and Partners

More than any other group, Foster and Partners sought to recreate the effect of the towers on the skyline. In addition to noting their desire to abide by the New Urbanist creed of a restored street grid, a vibrant people-centered community of small commerce and culture, and sustainability and security, this team emphatically stated that the "iconic skyline must be reassembled." Their design called for two separate towers that came together at three points, where they "kiss . . . creating public observation platforms, exhibits, cafes and other amenities." Presaging the memorial that would ultimately be built on the site, Foster and Partners also imagined a park with twin-tower-shaped voids (although their version possessed walls that rose a few stories out of the ground) to serve as a memorial.[88]

"The Dream Team"

Richard Meier, Peter Eisenman, Charles Gwathmey, and Steven Holl, on the other hand, harkened back to a previous era. "In the tradition of Rockefeller Center and Union Square," they announced, "we propose to build a great public space for New York City at the World Trade Center site. We call this Memorial Square." This was, of course, the same name as one of the doomed BBB designs, but the Dream Team appeared not to have made the connection. Particularly important to this team was the existence of multiple sites for memorialization at ground level. Some would be predefined by the team ("To the west, two glass-bottom reflecting pools demarcate the footprints of the former World Trade Center towers. Beneath them, the volumes of the footprints become sites for memorial rooms lit from above. The pools overlook two memorial groves of trees, planted to mark the final shadows cast by the towers moments before each fell"), and others would be subject of a design competition, particularly those that spread out from the sixteen-acre site into lower Manhattan. Key to this design was the latticed structure of the two buildings proposed at the site (which were meant to resemble interlocked fingers, but looked more like a hash tag with too many lines), which, because of their large horizontal space above ground, left twelve-and-a-half acres of the sixteen–acre site open for noncommercial development.[89]

The THINK Team

Rafael Viñoly, Frederic Schwartz, and their team sought to focus on the global dimensions of the World Trade Center site as well as the local. "Ground Zero should emerge from this tragedy" they argued "as the first truly Global Center, a place where people can gather to celebrate cultural diversity in peaceful and productive coexistence. Finding the proper balance between the two main objectives of the project—Remembrance and Redevelopment—depends on the way in which investment in the public infrastructure contributes to the Renewal of Lower Manhattan." In their vision, "The World Trade Center is reborn as the World Cultural Center. Built above and around the footprints of the World Trade Center towers, but without touching them, two open latticework structures create a 'site' for development of the Towers of Culture." These towers would reproduce the spatial look and feel of the twin towers and would include spaces for the memorial, a museum, a conference center, and a performing arts center nestled at various heights within the scaffolding of the towers.[90]

United Architects

United Architects chose a more mediaeval religious metaphor to explain their design. They proposed "an interconnected series of five buildings that creates a cathedral-like enclosure across the entire 16-acre site. A vast public plaza and park is formed around the connected footprints by a protective ring of towers." The proposal continued: "Preserving the footprints of the World Trade Center, the memorial visitor descends seventy-five feet below ground along a spiral walkway to then look up through the footprints to the sky. Rather than looking down, the memorial directs visitors to look upward in remembrance. A Sky Memorial atop the first tower will allow visitors to complete the memorial pilgrimage by looking down over the hallowed ground where so many heroes lost their lives."[91]

Skidmore, Owings, and Merrill

The team led by Skidmore, Owings, and Merrill, focused on multiple strata of horizontal spaces, including a series of public gardens high in the sky, rather than conceiving of the site as a single sixteen-acre plot of land with tall buildings on it. "In our proposal," which they called a vertical city, "the legible icons are the striations of space, rather than commercial structures." Ultimately, this team sought to simultaneously support the layered needs of the city and the public: public space, cultural space, and commercial space. It also called for a competition to design the memorial that would take shape at the site, but gave little indication of what this aspect of the project might look or feel like.[92]

Peterson/Littenberg Architecture and Urban Design

Finally, Peterson/Littenberg chose to focus on the concept of a public garden "whose shape and geometry are generated by the WTC tower footprints. This garden is a walled enclosure, quiet and contemplative, a place of allegory, historical remembrance, symbolism and repose. Within the garden and at other places throughout the district will be sites for an international memorial competition." On the fringes of this garden would be two new towers that, while not resembling the old towers, directly gestured to their memory.[93]

Design Selection

In closing out the event, LMDC chairman Whitehead thanked all of the architects for their participation and told the audience that over the next several weeks, the LMDC and the PA would evaluate each of the designs and put together a master plan based on the best elements of all of the proposals.[94] During this period, tens of thousands of visitors came to the Winter Garden to view the plans, and the LMDC held a massive public outreach campaign to inform the general public, and the relatives of victims, about the various site plans. The LMDC also convened a panel of experts to evaluate the plans on a variety of factors, including the extent to which the design provided a suitable memorial space; the extent to which it met the program elements of the site (e.g., adequate office, retail, and hotel square footage); whether it provided a workable street grid; whether it would transform the area into a "24/7" mixed-use neighborhood connected with surrounding neighborhoods; whether it provided a suitable space for phased, private development; cost, environmental and engineering issues, and the extent to which any problems found in the designs could be resolved efficiently and effectively.[95]

Victims' families also weighed in on the designs. The Coalition of 9/11 Families strongly supported Daniel Libeskind's plan because he integrated a memorial to the victims of the attacks into his design and sought preserve the slurry walls and the bathtub area (the underground excavation designed to keep water out of the World Trade Center site). This was precisely the area that these families considered sacred. Additionally, Libeskind's architectural and spoken rhetoric tended to remember the history of the site, and particularly the attacks, more than other plans. More generally, though, they reiterated their desire that the memorial planning process take place before the site design process and that the families be more involved in the entire process.[96]

On January 13, 2003, New York New Visions released its evaluation of the various site plans. NYNV noted that none of the designs were perfect, but that they at least represented bold and insightful thinking. In keeping with the spirit of the original Innovative Design Study, the NYNV committee did not seek to pick a single winner from the designs on display. Rather it chose to highlight the strongest ideas to emerge from the six teams in light of the criteria laid out in the RFQ. Indeed, one of NYNV's strongest statements was that "this process must be seen as a competition of ideas, not designers."[97]

In the context of evaluating the strengths and weaknesses of each proposal, and what such an analysis contributed to the overall understanding of the redevelopment of the site, NYNV highlighted what it considered to be the main issues: "The Lower Manhattan real-estate market is in both a recession and a restructuring; the Port Authority, which owns the site, is preparing its own plan; a private developer holds title to the lease; and funding is a complicated maze of agency allocations, tax credits, insurance proceeds, and miscellaneous grants. The crucial memorial design competition has not yet taken place. The public has not yet been involved in a clearly organized review process."[98] NYNV, therefore, called on the LMDC to exert "leadership and direction" in the development of lower Manhattan.[99]

In keeping with this lack of clarity over the process, on February 4, 2003 the LMDC announced that it was abandoning its mix-and-match planning strategy. Instead, two of the six teams were chosen to go head-to-head in a newly conceived final round: the THINK team and Daniel Libeskind.[100] Both were asked to make modifications to their plans in order to assuage concerns of the LMDC and PA, and other stakeholders. Focusing primarily on engineering issues, THINK reduced the size and diameter of its proposed towers in order to cut down on costs and altered their foundations so they didn't interfere with transit infrastructure. Further, the proposed reflecting pool and bridges were replaced by paved open spaces that could be accessed directly from street level. THINK also made changes to the street grid and pedestrian throughways.

The most significant modification made by Daniel Libeskind was a change in the slurry wall design. To begin with, it was covered up and no longer left open to the sky. Further, rather than exposing the slurry wall down to bedrock, the design element which so pleased the coalition families, Libeskind reduced the exposure to a mere thirty feet below ground in most places, in order accommodate transportation infrastructure that was being planned by the Port Authority for the site—specifically the PATH train and a bus depot to help handle the millions of tourists who would visit in a way that did not clog surrounding city streets.[101] Although there would be the opportunity to descend to bedrock in a part of the site, the entire bathtub area would no longer be preserved as sacred ground. Other changes included the morphing of the hanging garden of the proposed tower into a TV antenna-like structure and the reorientation of the proposed railroad station so that it opened onto the main square of the site and captured the "Wedge of Light" that would appear every September 11 at the time that the attacks occurred.[102]

The architecture community infighting and public relations efforts that ensued in the battle between these two master planners have been told elsewhere—on the Internet, in the mass media, and in several books on the subject.[103] Suffice it to say that the war of words in public and private was nasty as each group sought to solidify its position and as a cast of elite architects grumbled that they weren't chosen, and criticized the work and personality of those who had.

In the run-up to the final decision, no clear favorite emerged. The LMDC board and Mayor Bloomberg's point man on the redevelopment of the site, Daniel Doctoroff, favored the THINK plan, while Mayor Bloomberg, Governor Pataki, and the public seemed to favor Libeskind's plan. The Port Authority also seemed to be on Libeskind's side, particularly since he had been so accommodating to their infrastructure needs. Larry Silverstein, however, didn't like either of the designs, and still wanted to use his own architect, David Childs of SOM, to design the centerpiece of the five buildings on the site—WTC 1.

Given the political power and financial control that Governor Pataki held, it is no surprise that his influence won out. On February 27, 2003, he and Mayor Bloomberg announced that Studio Daniel Libeskind had been selected to plan the redevelopment of the World Trade Center site. To make the decision seem as legitimate as possible, the major stakeholders came together to praise the democratic nature of the process. They also noted that the decision to go with Libeskind was a milestone in the recovery of lower Manhattan. All parties, however, took pains to note that the master plan was really just a starting point, and that much remained to be done to come up with a practical, workable design for the site. In a prepared statement, Larry Silverstein wrote, "Our company eagerly looks forward to working closely with Studio Daniel Libeskind on developing their site plan."[104]

Families React

In a meeting with Libeskind on March 17, 2003, coalition families expressed their disappointment that he had made changes to suit the powerful interests shaping the World Trade Center redevelopment. They explained that what they had most appreciated about his initial design was that it preserved the bathtub down to bedrock as a memorial space, free of development. Libeskind said that in his view, the slurry wall was the element of the site that most needed to be preserved, not the bathtub down to bedrock. For the coalition, though, the slurry wall was "simply the demarcation where the greatest concentration of remains were found and where the families believed the Memorial should be built to bedrock."[105]

In an article in the coalition's newsletter, Beverly Eckert said it had become clear during the March meeting that the redevelopment of the site would favor interests other than their own.[106] In some ways, their pessimism was prescient. The reality, though, was that with so many interests at stake, no one's could be adequately met without trampling on the views and desires of other stakeholders. Even Larry Silverstein, whose interests were protected by force of contract and political will, would not escape a long, damaging battle to achieve his goals.

Over the next decade, pundits, critics, and the media would feast on the intrigue, expense, and infighting that would take place at the World Trade Center. Every news story seemed to confirm that the families had little more than symbolic influence in the redevelopment efforts. Many families felt continuously revictimized by the inability of the powers-that-be to settle their differences in a dignified matter. Others learned to take part in these battles and exert their will on the process, albeit in a compromised fashion. But for now, at least, the master plan search process had come to a conclusion. This milestone meant that a new chapter could begin: the selection of a memorial for the victims, survivors, and heroes of the September 11 attacks.

5

Memorial

In *This Republic of Suffering: Death and the American Civil War*, Drew Gilpin Faust explains that "it is work to deal with the dead, to remove them in the literal sense . . . and to remove them in a more figurative sense."[1] This sentiment is as true for the World Trade Center attacks as it was for the Civil War. Remains of the victims had to be recovered, identified, and returned to families or placed into storage, or preserved for potential future identification. Just as importantly, the land on which the deaths took place had to be cleaned up. In addition, as Faust notes, families and communities affected by mass death "must repair the rent in the domestic and social fabric, and societies, nations, and cultures must work to understand and explain unfathomable loss."[2]

From the moment the twin towers fell, it was clear that the attacks on the World Trade Center would be seen as a transformational moment in U.S. history—like Gettysburg, Pearl Harbor, or the assassination of JFK. Americans, or at least those who were old enough to understand the import of the attacks, would come to see the world in two distinct time periods: before 9/11, and after. It is unsurprising that Americans almost immediately sought to find a broader meaning in the events. Was 9/11 a cowardly attack carried out by radical extremists who hated freedom and democracy, as President Bush told the country? Or was 9/11 a response to America's long history of economic and military exploitation of the less powerful around the world—the "chickens coming home to roost," as controversial scholar and Native American activist Ward Churchill infamously put it in a September 2001 essay?[3]

For most families of the missing, though, the days and weeks after the attacks were filled with personal shock and frantic searching, which made efforts to contextualize the deaths of their loved ones ring hollow and often seem callous. Families would ultimately, however, be forced to confront the larger meaning of the deaths as the initial grief gave way to social, political, and cultural acrimony over the best way to remember and memorialize them.

Missing Persons Posters

It has been widely recognized that the first memorial to the victims of the World Trade Center attacks were not intended to memorialize them at all—they were the missing persons posters that blanketed lower Manhattan, and the entire region, in the days after the attacks when the death toll was believed to be as high as 6,000.[4] These posters were simple and to the point: a photograph of the missing person (often with family, on vacation, or at a party), some basic details about his or her appearance—both banal (height, weight, eye color) and more intimate (tattoos, piercings, scars, and other distinguishing characteristics previously known to few people), information about where the missing person worked and what he or she was believed to be wearing that day, and contact information for the person doing the searching. Many posters also contained special pleas for attention—perhaps the missing person was pregnant, had young children, or was simply a kind and gentle soul—as if to acknowledge the overwhelming nature of the loss.

Kinkos, the nearly ubiquitous copy shop, offered free reproduction for these posters, and soon they covered walls, hospital entrances, store windows, mailboxes, phone poles, subway stations, and any other spaces that might be seen by passersby. Then, as journalist Amy Waldman noted a few weeks after September 11, "manufactured in hope, the fliers now have transmuted into memorial." She quotes James Saunders, the public affairs director at Bellevue Hospital Center, who said that the flyers taped to the concrete barrier in front of the hospital (which had been rechristened the Wall of Prayers in the days after the attacks) had "become for us here, and many who walk past, much like the Vietnam Veterans Memorial. You walk past and see their faces and acknowledge that these were lives."[5] A similar "Wall of Hope and Remembrance" emerged at nearby St. Vincent's Hospital.

The flyers, Waldman lamented, "now simply present physical proof, a public statement, that the person lost was loved. They impress individuality on an event so gross in scale that an abyss of anonymity threatens the dead."[6] As time went on, the posters became de facto memorials and achieved a status as icons of the event. In the weeks and months after 9/11, they were photographed and collected, and have since gone on display in many exhibitions in New York and around the country. Indeed, many families and friends would return to the posters to write their goodbyes to their loved ones when their remains were found or it became clear that they were not coming home.

Immediate Memorials

Memorials to the missing and dead sprang up almost immediately in the New York metropolitan area and elsewhere around the country and the world.[7] Union Square, which had long been a gathering place for social and political activism and was one of the nearest large public spaces that had not been affected by the shutdown of lower Manhattan, became the epicenter of public grief. People gathered there in the hours and days after the attacks to leave missing persons posters, flowers, candles, American flags, numerous replicas of the twin towers, and other items to remember and honor the victims of the attacks. They played music, talked with friends and strangers, and contemplated what life would be like moving forward.[8] Visitors had many reasons for coming to these early memorials—to seek solace, to be with people who understood what they were feeling, to be part of the collective rather than an island unto themselves, to feel hopeful for the future, to begin to make sense of what had happened, and to put their minds at ease.

Recognizing that people had more than just an urge to be together in frightening and uncertain times, a group of New York University students began rolling out scrolls of butcher paper along the paths of Union Square and leaving markers so people could communicate their fears and feelings. Once a scroll was complete, it was hung on a fence or moved to the side of the park so that a new scroll could be laid out. More than 200 of these scrolls were eventually donated to the New York State Museum, and they were ultimately catalogued and put on display at Siena College for the tenth anniversary of the attacks.[9] These scrolls contain messages of hope, despair, loss, grief, pain, confusion, peace, and worry. Some writers included a favorite biblical verse, poem, or saying, while others drew pictures of the towers as they stood before September 11.

In addition to the scrolls, individuals and small groups created shrines in various parts of the park. These shrines tended to be patriotic in nature—asking America to be strong and resilient, letting the people of New York know the nation was with them, and expressing sorrow for the lives lost in the attacks. Most called for tolerance and peace, including a peace sign painted on the hindquarters of George Washington's horse, a peace flag placed in Washington's hand in a statue in the park, and signs like the one that stated "SAFETY BEGINS AT HOME: Protect your Arab & Muslim Neighbors."[10] Similar memorials sprang up on a smaller scale in places like Washington Square Park, the Hoboken Waterfront, the Brooklyn Heights Promenade, and elsewhere around the region.

St. Paul's Chapel, just a block away from the World Trade Center site, became another spontaneous memorial location when public access to it was restored a few weeks after the attacks.[11] Like Union Square, the chapel was inundated with flowers, candles, American flags, and other symbols of grief and solidarity, especially teddy bears and foreign flags. But it also became a place to mourn the uniformed service personnel who lost their lives in the response to the attacks and to celebrate those who were working to recover the remains of the dead. The church hung huge canvas sheets from their fence so that people could write on them. People left messages of thanks to the firefighters, EMS workers, and police who were working at the site, and memorials to those who were missing or already found dead. Mementos like hats, t-shirts, and sweatshirts were left as homages from around the country.[12] Similar memorials emerged in front of firehouses and police precinct stations around the region.

People also placed shrines, banners, and memorials to the dead on the security fence at Liberty Plaza, which was set up to cordon off Ground Zero from the rest of lower Manhattan. They did the same at the Fulton Street viewing platform, which the Lower Manhattan Development Corporation built in December 2001 to create a safe place for the public to see what was going on in the Pit. More privately, family members decorated the LMDC's family room with missing posters, along with personal items belonging to their loved ones, flowers, teddy bears, handmade shrines, and other forms of remembrance.[13]

The pervasiveness of these acts, along with hundreds of 9/11-related Internet memorials, highlights people's need to mourn the events of 9/11 and to share their grief, anger, and fear with others. Indeed, in what could be seen as another form of memorialization, Laura Kurgan, an architect and designer, produced a free, fold-out map of the Ground Zero area that explained to visitors what they were seeing, where the World Trade Center buildings once stood, where temporary memorials could be found, and which streets were open and which were closed. The map provided photographs of memorial sites and encouraged users to make their own contribution to the memorialization process. In March 2002, Kurgan wrote:

> The map thus serves as a document of a place that is changing, of a line in the city that will soon disappear as the debris is removed. Presently, the site is a mass grave in the process of being disassembled and removed. This is a temporary condition that will soon disappear into a complex political debate over different proposals for rebuilding and memorializing. The map exposes this raw aspect of the site, the mass grave, as the one which should be memorialized. To emphasize this, we have included photographs of the temporary memorials contiguous to the path and of the green fences which have become home to innumerable flowers and letters and, of course, teddy bears.[14]

Kurgan's statement offers a clear vision of the relationship between the site as it existed in the present and what ought to be memorialized in the long term. It also highlights the emerging conversation about more official and more permanent memorials to the victims and events of 9/11. Her choice of words to describe these discussions—complex and political—could not have been more prescient or accurate.

Toward a Permanent Memorial

In an effort to start the memorialization process, New York New Visions held a series of conversations with key stakeholders in November and December 2001. These events brought families of victims, rescue workers, downtown residents and business people, property owners, civil leaders, students and educators, and the lower Manhattan financial community together in small groups to talk about their concerns and hopes for any memorial that would eventually be created. Families of victims who participated were adamant that a memorial needed to be built to the victims of the 9/11 attacks and that this memorial should be dedicated to those who lost their lives on September 11 whether they were killed at the World Trade Center, in the plane crash near Shanksville, Pennsylvania, or in the attack on the Pentagon. The memorial would have special resonance for the families of the World Trade Center attacks, however, because for those who never received remains, it would be the place where they go to mourn the loss of their loved one.

Relatives debated the merits of creating a museum on the site, but were clear that if one were built, it should focus on the victims of the attacks and not tell the story of the perpetrators. In general, this would come to be a steadfast position of many, but not all, family members—that the actions of the terrorists were senseless and inexplicable, and that any attempt to provide a logical explanation of their ideology or tactics would be tantamount to dignifying them and their cause.[15] This view would come to the fore in debates over whether or not to locate the International Freedom Center on the memorial space at the World Trade Center site, and whether or not to show photographs of the terrorists in the 9/11 museum that was eventually planned for the site.

In the family focus group, one participant voiced concerns about the takeover of the memorial process by rescue workers, who were being immortalized in the media and by politicians. This statement led to a conversation about the relationship between civilian and uniformed service personnel in any planned memorial, leading to a consensus that all victims should be treated equally.[16] Civilian families would make this demand numerous times over the next several years, noting that many civilian victims acted unselfishly and heroically after the attacks.

It is important to note that representatives of rescue workers, while they did want the bravery of firefighters to be highlighted, were not calling for separate or preferential treatment. Indeed, one firefighter explicitly stated that he felt that his fellow firefighters were getting too much press attention, making it more difficult to hear civilian voices. Another noted that the World Trade Center was a microcosm of New York—with rich, middle class, poor, and even homeless people interacting on a daily basis. Thus, he felt that the memorial should reflect this diversity and remember everyone who died there.[17] While debates would rage about whether and how to denote rescue workers at the memorial, not all uniformed service personnel believed this was the right thing to do.

Rescue workers and civilians found much else to agree upon—particularly the notion that the site itself must include a significant memorial. As the rescue workers noted, it was a "graveyard" for so many people, and the families of victims noted it would be the only place that people without remains could go to mourn their loved ones.[18] Both civilian and uniform families also agreed that the Port Authority and Larry Silverstein were making decisions without their input. Families felt that they needed to organize in order to compete on an even playing field. Rescue workers, while less concerned that their own voices were not being heard, expressed concern that the local residential community most affected by the attacks was not being adequately represented in the decision making.[19]

Residents and business people of lower Manhattan also stressed the need for a memorial, but their vision was less focused on the dead. In addition to memorializing the tremendous loss of life, the memorial ought to remember the effects of the attacks on the local community and the economy—both in New York and nationally. They were concerned that rather than focus solely on violence and death, the memorial should promote peace and social harmony. Finally, they stated their objections to the term "graveyard" and hoped that another metaphor for the site would emerge.[20]

Temporary Memorials

It was clear to all stakeholders that the creation of a permanent memorial would take time, particularly if it was to be chosen through an international design competition, for which many groups and stakeholders were already advocating. In the meantime, the tremendous interest in the site, and the continued struggle to come to terms with the effects and meaning of the attacks, meant that the need for temporary memorials would have to be addressed. Among the first efforts to honor those who died in the attacks was the decision by the Mayor's Office to turn Fritz de Koenig's *Sphere for Plaza Fountain,* the bronze sculpture that was centrally located in the plaza of the World Trade Center site, into a memorial in a small park at the southern terminus of Greenwich Street, adjacent to the World Trade Center site. Koenig's *Sphere,* which was heavily damaged but not destroyed by the collapse of the towers, quickly became an icon of the attacks.

This decision created a controversy when many residents of lower Manhattan, led by Community Board #1 chair Madeleine Wils, protested that the temporary memorial did not belong in a residential area.[21] They were concerned that the memorial would generate too much traffic for the area and that it would serve as a depressing reminder of September 11 for a community struggling to return to some sense of normality.[22] After threats of civil disobedience by residents, the plan to place *Sphere* adjacent to the site was scrapped and it was instead installed, on March 11, 2002, in Battery City Park alongside an eternal flame in memory of the victims of the attacks.[23]

Families found the decision to use Koenig's *Sphere* problematic, but not for symbolic or any other material reasons. Rather, they objected to the fact that they were not asked to participate in the decision, but were simply informed by Christy Ferer, Mayor Bloomberg's family liaison, that the sculpture would be used as a memorial and invited to the dedication ceremony.[24] Many activist families would continue to complain about a perceived lack of appropriate consultation throughout the redevelopment and memorialization process.

The other major act of official temporary commemoration was the "Tribute in Light," in which high-wattage search lights were configured to reproduce the shape and location of the twin towers in the nighttime skyline from March 11 to April 14, 2002. The display, both simple and elegant, was independently conceived by several artists and designers and was initially operated by the Metropolitan Arts Society through funding from the LMDC. This tribute has been popular with all constituencies and has been repeated every year on the evening of September 11. Indeed, it has become an iconic image of post-9/11 coverage of the attacks, appearing on the covers of *Time*, *Life*, *Newsweek*, and many other magazines and newspapers after its debut in 2002. In 2012, the National 9/11 Memorial and Museum Foundation agreed to take over and keep the tradition alive for the foreseeable future.

While both temporary memorials created to mark the six-month anniversary elicited a great deal of positive media attention, and were applauded by the public, for many family members they remained a reminder of how much work still needed to be done to create a meaningful and dignified permanent memorial.[25] The most important concern was the extent to which the permanent memorial would be similar to the temporary memorials, or would more explicitly celebrate the victims as individuals. There was even discussion about whether the memorial ought to serve as a final resting place for unidentified and unclaimed remains. As Monika Iken said in 2002, "We have nowhere to go. You know, we have to grieve in public.

I don't have an urn. I don't have a coffin. I have nothing. I have nowhere to go to mourn my husband and most of the families have nowhere to go as well. . . . Going there to this day, it's hard to explain, but I still feel him. I feel the energy of the souls. I feel their crying. They're not at rest. It's not a peaceful place at all. Now I don't see the towers even in my mind anymore. All I see is a graveyard without tombstones."[26]

Crafting a Mission Statement

Aware of the clear need to make the Families Advisory Council (FAC) central to the memorial planning process, the LMDC gave the FAC the opportunity to make recommendations about the design process and the themes and messages family members wanted the memorial to portray to visitors. The FAC was divided into two groups to help facilitate this process. The program subcommittee met for the first time on June 4, 2002.

From the beginning, it was clear that the LMDC hoped to moderate the influence of the families that were calling for the preservation of all sixteen acres of the site and demanding that footprints of the towers remain inviolable. They did so to a large extent by putting Thomas Rogér and Nikki Stern, family members who were not part of the activist Coalition of 9/11 Families, in charge of the process. Rogér, a building company executive and the father of Jean Rogér, a twenty-four-year-old flight attendant who was killed when American Airlines Flight 11 crashed into the North Tower, did not share the physical connection to the WTC site that many of the coalition families, whose loved ones worked in the towers or were uniformed service personnel responding to the emergency there, felt. His organization, Families of 9/11, was briefly a part of the coalition, but dropped out when it was clear that the group was going to take oppositional stances against the LMDC and the city. He simply could not understand the idea that the footprints of the tower were sacred or that Ground Zero was tantamount to a cemetery.[27] In his view, such a claim was a logical fallacy. "The fact is," he told journalist Jennifer Senior, "the remains of 1,500 people have not been recovered, so in some people's minds, they're still there [but] they're not. . . . [They] were scattered everywhere." He also felt it was also a capitulation to the terrorists themselves, who were "the only people who chose to die there." In Rogér's view, the families would have had a much easier time if they chose another place to associate with their loved ones—a location that could bring solace rather than pain. "I mean, my daughter—her graveyard, in our minds, is not going to be there. Symbolically, we've already committed her remains to a place on Lake Erie, where my family has a home. As far as we're concerned, that's where she is."[28] Interestingly, Senior noted that the remains he mentioned were actually World Trade Center debris that had been sanctified by the City and NYPD. This suggests that Rogér, like Mayor Bloomberg, chose to remember his daughter through his memories of her rather than through the remnants of her body.

Stern, who lost her husband James E. Potorti, an executive at the securities firm Marsh & McLennan, was more agnostic about the need to preserve the footprints and was skeptical, but not as dismissive as Rogér, of the notion that the World Trade Center site and the footprints were "sacred ground."[29] It is important to note that, over time, Stern would become a critic of the behavior of some family groups and individuals who were most active in debates about the memorial.[30]

At the June 4 meeting, Stern began by setting out the parameters of the debate. She told the subcommittee that the mission statement should answer four concrete questions: who is the memorial for, what does it commemorate, what will it say to the future, and what will it include? The subcommittee had a wide-ranging discussion of these questions before Rogér presented the first draft of a "September 11 Memorial Mission Statement" that he had written based on his reading of the Oklahoma City Memorial mission statement. Stern then collected comments from other subcommittee members in an effort to rework it.[31]

Over the next several weeks, the program subcommittee met to discuss and debate the draft mission statement. By mid-August, they had produced a working draft, which was approved by the full FAC and then mailed to all families of victims for feedback and presented at the Listening to the City events in July 2002.[32] This preliminary draft and related comments became the starting point for the official LMDC Memorial Mission Statement Drafting Committee and Memorial Program Drafting Committee, both of which were convened in November 2002.[33] The goal of the Mission Statement Drafting Committee was obviously to get the wording of the mission statement so clear that it could guide the design competition over the next few years, and the operation of the memorial in perpetuity. The goal of the Program Drafting Committee was to establish: (1) guiding principles that laid out the "aspirations that must be embodied within and conveyed through the memorial regardless of the various interpretations to which they will ultimately be subject;" and (2) a set of program elements intended to "provide memorial designers with a list of specific elements that should be physically included in the memorial, without prescribing how or inhibiting creativity."[34]

Both committees were diverse in representation, with three family members being involved in this process: Christy Ferer (the Mayor's family liaison), Paula Grant Berry, and Tom Rogér, but only one active member of the Coalition of 9/11 Families, Kathy Ashton from Give Your Voice, was included in either committee.[35] This prompted an outcry from the coalition, which felt that only family members who were on board with Mayor Bloomberg's vision for the World Trade Center site were chosen.[36] A few days later, Monica Iken was added to the Memorial Program Drafting Committee.[37] According to the *New York Times*, Governor Pataki had been dismayed at the public criticism of the LMDC memorial process and had urged its leaders to add her to the committee. Adding weight to this claim, Iken publically thanked the governor for "understanding the importance of the families being part of this process."[38]

As the committees were being formed, the LMDC named Anita Contini, who had spent time in the corporate world and was an art curator and public arts advocate, vice president for memorial, cultural, and civic programs, with a mandate to manage the memorial development process. One of Contini's first major actions was to coordinate visits by LMDC staff, board members, and FAC members to memorials around New York City and to the major memorial sites in Washington, DC. They also visited the Pentagon, the United Flight 93 crash site in Shanksville, Pennsylvania, and the National Civil Rights Memorial and Rosa Parks Museum in Montgomery, Alabama.

The LMDC then offered a presentation to the public that highlighted what it learned from these visits. This presentation provides a window into the minds of LMDC officials at the beginning of the memorial design process.[39] To begin, the LMDC highlighted the links between the three sites affected on 9/11, demonstrating that LMDC officials viewed the World Trade Center memorial as representing the totality of 9/11 and not just the events that took place in lower Manhattan. The LMDC also highlighted the value of open competitions for the memorial design, simple mission statements to guide competitors, and professional juries. The LMDC recognized (using the Vietnam Veterans Memorial and the United States Holocaust Memorial Museum as examples) that controversy and debate was to be expected in the design process. The visit to Arlington National Cemetery provided the opportunity to reflect on how unidentified and unclaimed human remains might be dealt with. The presentation noted that "unidentifiable remains of those who died in that attack on the Pentagon have been cremated and interred in a casket in section 64 of Arlington National Cemetery, which received ceremonial burial on September 12, 2002." The site was marked by a large headstone, which recognized "all 184 people lost in the attack."[40] The presentation also noted that the cemetery was home to the Tomb of the Unknowns, which contains the remains of unidentified soldiers from World War I, World War II, and the Korean War. Finally, the LMDC promoted harmonious design in which the architecture itself would be part of the memorial. It did not take a position on grandness vs. intimacy, but argued in favor of a memorial that inspired visitors, would have continued relevance for future generations, and functioned as a living memorial.[41]

With this background trip complete, the committees began crafting the mission statement and memorial program. Around the same time, New York City Council member Alan Gerson asked New York New Visions and the city's American Institute of Architects chapter to bring together a group of relatives of victims and residents of lower Manhattan over three sessions to determine how much common ground existed between the two groups on the issue of the memorial. The media seemed to delight in portraying the relationship between families and residents in a negative light, and Gerson wanted to know on what principles they actually might be able to agree.[42] Family members and residents were able to come to consensus on several major points, most importantly that the entire sixteen-acre site be considered hallowed ground; the bathtub area (as defined by the Coalition of 9/11 Families) be considered "sacred space"; the area around the bathtub be devoted to a "living memorial" highlighting the resilience of the lower Manhattan community and humanity in general; vehicular traffic be limited throughout the site but pedestrian connectivity encouraged; and short-term solutions to memorial and community infrastructure be developed while long-term solutions were being planned. NYNV offered these points of agreement as "Four Principles for the Memorial Program."[43]

Much of the intent and sentiment of the preliminary FAC draft survived the versions produced by the official committee without significant revision—most notably the importance of remembering and honoring those who died as individuals; acknowledging survivors and all who were part of the initial response and rescue efforts, as well as those who participated in the recovery efforts; conveying the "historical significance and global impact of September 11, 2001"; "inspiring and engaging people to learn more about the events and impact" of the attacks of 1993 and 2001; recognizing the positive aspects of the response to the tragedy, in which the nation and world came together to denounce violence and hatred; and providing separate areas for families and loved ones to reflect on and remember the victims of the attacks.[44]

Of the changes that took place between the FAC subcommittee preliminary draft and the final Memorial Mission Statement and Program, some were subtle and primarily semantic, while others had a profound impact on the memorial that was ultimately built. The victims of the terrorist attacks in 1993 and 2001 were described in terms of "innocent lives lost" in the preliminary FAC version, "killed in the horrific attacks" in the first draft of the official Mission Statement committees, and "murdered by terrorists" in the final Mission Statement and Program. The Mission Statement Drafting Committee spent many hours debating word choice and noted that the decisions made "best reflected the intentions of the committee."[45] This phrasing was also in accord with the feedback received from families and the public during the summer 2002 outreach campaign.[46]

Further, the preliminary draft went to great lengths to mandate the preservation of the footprints as sacred ground and that only memorial-related development should occur within these spaces, while the draft statement and final version only mandated that the memorial "respect this place made sacred through tragic loss" and "make visible" the footprints of the original towers.[47] Further, both of the later versions only stated that the memorial "may include" surviving original elements and preservation of existing elements of the site, not that they must do so. In another subtle but important shift, while the preliminary FAC draft stated that the entire sixteen-acre site was sacred and ought to be considered "as the final resting-place for our loved ones" and called for "an area for internment of unidentified remains," this broad statement morphed into a more limited programmatic requirement for "a separate, accessible space to serve as a final resting place for unidentified remains from the World Trade Center site." In other words, the entire site was not considered to be the final resting place for the victims, only the location for such a final resting place. The memorial competition guidelines did not specify any particular requirements for this space other than to show where such a space would be and how one would get in and out of it. The competition guidelines allocated approximately

2,500 square feet of enclosed space for the remains, but told designers that the design for the interior of this space did not need to be provided in their submission.[48]

Perhaps the most surprising feature of the entire effort to create a memorial mission statement and program was the lack of a budget discussion. Indeed, the final version of the mission statement and program noted "the budget for the memorial will depend upon memorial design."[49] It was only in the second stage of the competition, when the field was narrowed down to a few finalists, that cost estimates were required. This failure to integrate budgetary matters into the memorial program would haunt the LMDC for more than a decade.

Another surprise was that nowhere in the memorial program was there a requirement that the memorial design conform to the selected master plan for the site. While the memorial competition guidelines presented the basic aspects of Daniel Libeskind's plan and asked competitors to be "sensitive to his vision," the document also clearly stated in bold letters that proposals "to exceed the illustrated memorial site boundaries may be considered by the jury if, in collaboration with the LMDC, they are deemed feasible and consistent with the site plan objectives."[50] According to juror James Young, this language was inserted at the request of several jury members to ensure that competitors were not stifled by what appeared to be a memorial already over-determined by Libeskind's vision and the requirements of the site.[51]

Concerned that this language would be too subtle for many competitors (and clearly not being infatuated with Libeskind's design), the jury more explicitly encouraged designers to push beyond the limit in their public statement on April 28, telling them that they would not be faulted for doing so. Although the LMDC was unaware that the jury would make such a statement, LMDC president Kevin Rampe publicly praised the jury for asserting its independence. Whether this was a genuine expression of agreement, an effort to further distance itself from any decision the jury would ultimately make (knowing that it would be controversial and unpopular for many), or just a way to save face, is not clear.[52]

But entrants to the competition were expected to conform their design to the guideline that the "memorial site ground plan is a concrete deck located approximately 30 feet below street level and approximately 40 feet above bedrock. Memorial designs should not extend below this plane except at the northwest corner of the memorial site where the exposed slurry wall extends down to bedrock."[53] This dictate, of course, directly contradicted the view of sacred ground so tenaciously held by the Coalition of 9/11 Families. Indeed, these families wished that the LMDC had included a requirement that the North footprint down to bedrock be included as a program element of the memorial, with as much of the South footprint being preserved as possible. The LMDC hedged, telling the Family Advisory Council that did not want to place any such design burden on competitors.[54]

The Competition and the Jury

Almost from the beginning of the planning process, LMDC officials settled on a juried competition open to anyone in the world above the age of eighteen. Both the Vietnam Memorial and the Oklahoma City memorial were chosen this way, so there was little impetus to try any other route. An open competition would allow the global public to participate in the memorial process in a way that promoted transparency and also healing, and might even lead to the discovery of the next Maya Lin—an inspired, but unknown designer whose vision could become a timeless icon.[55] The only caveat was that "if the finalist or team member is not a professional architect, the finalist may be required to associate with an architectural firm or other professional. LMDC may require the finalist to associate with an architectural firm selected by both the finalist and the LMDC."[56]

The rules of the competition were relatively simple: register an intent to submit a design by May 30, 2003, and submit a proposal as a poster board that meets specific guidelines, along with a $25 entry fee, by June 30, 2003. Submissions would be processed in July to ensure that they met the requirements and rules of the competition, and the jury would have its first look at the anonymized entries in August. The LMDC hoped that the jury would be able to winnow the entries down to five finalists by September and then work with these individuals or teams to refine the designs based on site plan updates, overall aesthetics, and technical issues. The refined final designs would be presented to jurors in the fall; they would then undertake final deliberations and select a winner in October.

In addition to her role in the mission statement and memorial program and organizing the competition, Contini was also responsible for putting together the jury that would select a single design from the open competition. In many people's views, the jury was necessary to prevent design by committee and too much public input into the decision. The focus ought to be on design excellence and long-term value over political or economic expedience or short-term emotional resonance.[57]

According to architecture critic Paul Goldberger, given the intense politics surrounding the redevelopment of the site, "it was notable that there seemed to be little political pressure on the choice of a memorial design. Indeed, politicians seemed at pains to show how willing they were to isolate the memorial from the forces that they themselves had brought to bear on the rest of the process."[58] There were only two people on the jury with clear political connections—Patricia Harris, a deputy mayor, and Michael McKeon, Governor Pataki's former director of communications and former chief spokesperson.[59] While the exact calculus that Contini and her LMDC colleagues used to choose the jurors was never revealed, one juror, James Young, an academic expert on memorialization, explained that "chosen from all walks of life, but with very particular credentials, they were (the LMDC hoped) both representative enough and revered enough in their professional worlds to bring an unimpeachable authority to whatever design they would finally choose."[60]

By far the most well-known of the jurors was Maya Lin, whose abstract design for the Vietnam Veterans Memorial, conceived when she was an architecture student at Yale, was initially the subject of bitter controversy but had with time become an icon of design and perhaps the memorial by which all other American memorials are now compared. Lin also designed the Civil Rights Memorial in Montgomery, Alabama and was involved in numerous architectural projects in New York City and around the country. Other architects and artists included Enrique Norten (a professor of architecture at University of Pennsylvania), Martin Puryear (an internationally famous sculptor), and Michael Van Valkenburgh (a landscape architect who split his time between New York City and a faculty position at Harvard). The jury also included several prominent members of New York City's arts and cultural community: Susan K. Freedman (president of the Public Art Fund and a major supporter of the arts in New York City); Vartan Gregorian (president of the Carnegie Corporation of New York and former president of Brown University), Nancy Rosen (a longtime public art consultant), and Lowery Stokes Sims (executive director of the Studio Museum in Harlem). James Young (professor and chair of the Department of Judaic Studies and Near Eastern Studies at the University of Massachusetts) joined the jury as an expert in the memorialization process, with a particular focus on the Holocaust, and David Rockefeller (the philanthropist and businessman) served on the jury in an honorary capacity, as a prominent supporter of the arts in New York and as the force behind the original revitalization of lower Manhattan in the 1960s that led to the redevelopment of neighborhood that became the World Trade Center. Julie Menin, a longtime resident and business owner in lower Manhattan, represented the interests of those constituencies on the jury.[61]

Family Reaction to the Competition Format

The most striking aspect of the composition of the jury was the very limited representation of families. The sole family member on the jury was Paula Grant Berry, a publishing executive who lost her husband in the South Tower of the World Trade Center. With degrees from Harvard University and Columbia School of Business, she certainly had the credentials to serve on equal footing with the rest of the jurors, but many family members, particularly those from the Coalition of 9/11 Families, questioned her allegiance to their particular needs and interests.[62]

More generally, many family members felt that they were being marginalized by the jury process that gave them only one representative—and one that not all parties believed truly shared their goals.[63] They cited the Oklahoma City memorial as an example of a competition that provided extensive opportunities for family members to select a memorial design that was both true to the emotional needs of families and the community and excellence in design. Indeed, the contract with the architecture and design consultant hired to head the memorial selection process in Oklahoma City was terminated when it became clear that he did not think that families and survivors of the bombing should have a say in the selection process.[64] He was replaced with a team of three consultants who agreed to work closely with families and survivors during the judging of the design competition there. In the end, the competition guidelines stated that "the wishes of the Family/Survivors Liaison Subcommittee are to be given the greatest weight in the Memorial planning and development process" and a majority of the panel that selected the final design were family members or survivors (eight of fifteen).[65] Thus, in a sense, professional designers and architects served as advisors and were subordinate to families and survivors.

In New York, a very different system was put into place. The jury met with various stakeholder groups and the public in an effort to understand what they hoped the memorial would achieve, but were not bound in any way to give priority to their wishes.[66] The jury also did not engage with stakeholders, but rather listened passively to their testimony. In the strange logic of the LMDC, however, the interchanges that took place in its three "Public Perspectives Outreach Campaign" events were deemed a "dialogue." As acting LMDC president Kevin Rampe noted in his opening remarks at the "Joint Meeting of the Memorial Competition Jury and Families Advisory Council" on May 27, 2003, "We really want to have a dialogue that allows the jury to—they're not going to be answering your questions, but they are going to be listening in on a dialogue from the family members amongst themselves."[67] He went on to say that all questions posed by family members would be answered by the LMDC—at a later date. He also told participants the dialogue would focus solely on questions that the LMDC brought to the table, namely what should the memorial convey to future generations and what should the jury keep in mind as they moved forward with the evaluation process.

After extended introductions by the jurors, the meeting's moderator, Todd Jick, returned to the format of the evening: "You had a chance to listen to the jury. . . . But that's the last opportunity this evening you will have that chance. They have in turn now asked for you to speak and for me to play the role of . . . enabling that conversation to happen around the two questions which I think you have in front of you. . . . Because the way I look at it . . . is that we are going to have a conversation with a group listening in." He then told the family members not to try to "undo previous decisions," but rather to focus on the future.[68]

This attempt to control the conversation did not sit well with all of the family members in attendance. Jack Lynch, for one, demanded that the restrictive format be thrown out so that family members were allowed to speak their minds and could engage in a two-way dialogue with jury members. Marian Fontana stated that she felt that this attempt to control the conversation was another example of the disempowerment and marginalization of family members at the expense of other stakeholders like Larry Silverstein. Anthony Gardner complained that family members were being left out of the decision-making process and being relegated to an advisory role. Others, including Kathy Ashton, who lost her twenty-one-year-old son Tommy, did not endorse the process, but appreciated the opportunity to speak her mind to the jury.

Once conversation about the memorial got underway, participants focused on a variety of issues. Some, including Nikki Stern and Carol

Ashley, offered more general commentary about several topics, including the need to balance competing interests and the fact that many would be disappointed with the memorial that would eventually be built. Others focused specifically on elements or characteristics that they would like to see embodied in the memorial—especially the preservation of the entire bathtub area and the actual tower footprints from bedrock to street level. Jenny Farrell, the sister of twenty-six-year-old James Cartier, and Kathy Ashton asked jurors to adhere to the memorial mission statement and pick a design that did not create a hierarchy between rescue workers and civilian victims. Ashton told jurors that the effort by some in the uniformed service professional community to attain distinction for rescue workers at the site had been "hurtful, divisive, and judgmental."[69] Gardner asked that any museum that accompanies the memorial include no other events besides September 11 and the 1993 World Trade Center bombing.

Mary Fetchet, who lost her twenty-four-year-old son Brad, reiterated the statement that many of the families wanted more of a say in the memorialization process and asked the jurors to think of the families not solely as people to be heard, but as resources to be consulted. From the perspective of the memorial, she noted that Libeskind's original plan resonated with so many families because it preserved access to bedrock, slurry wall, and ramp, which many consider to be the final resting place of their loved ones. Access to this site was important because so many people would never receive remains to bury and any repository built there would likely hold at least some remains of a large percentage of victims. She further expressed her belief that survivors of the attacks and people who worked in the area, or who had a special connection to the towers, would also find solace at bedrock.

Fetchet also placed the memorial in the context of the forensic identification process. In her view, the jury needed to keep in mind the intense fragmentation of most victims and the fact that many families had been, or would be, notified multiple times about the identification of remains. The complexity of the identification effort and the fragmentation of bodies, she told jurors, "just validates that you don't really have a place other than the site."[70] More generally, Fetchet worried that the information that accompanies the identifications, or the lack of any information at all, traumatizes families and makes it difficult for them to think properly about memorialization. In Fetchet's view, the process was happening too quickly, making it hard for her to properly articulate her own needs or those of her surviving children. She closed by reminding the jurors of their power: "You can tell the character of a man by what he does for the man who can offer him nothing. And I think that you're in a position right now, we can only advise you, we can only be here to be a sounding board for you, but you are representing hundreds of thousands of people and the world is watching."[71]

Chris Burke, whose brother Tom died in the attacks, reaffirmed Fetchet's statement. He then told the jurors: "I am one of the 1,400 families that did not get a phone call [from the Office of the Chief Medical Examiner]. Terms like 'hallowed ground' and 'sacred ground' are so much more than terms to my family, to my brother's children." Although Burke recognized that lower Manhattan must return to life as a vital business district and residential neighborhood, he literally begged the jury not to forget that so many lives were lost at the site, "never to be regained, so many lives without even a proper burial. . . . I do not want to buy a frappacino [sic] on that site. This is my brother's and so many others' final resting place. It must be respected."[72]

Not all family members thought of the site as their loved one's gravesite. For Virginia Bauer, "When I go to the golf course, when I look at my home, when I look at my children, that's what makes me think of my husband." That said, she believed that, among active family members, she was unique in her views.[73] Suggesting that her views might have been more widespread than she had assumed, two other family members made similar comments. Elinore Hartz, whose husband John worked in the towers, said that "I went down to the World Trade Center about three weeks after. I remember looking at that pile of rubble and thinking my husband is not there. And I was very annoyed that they were asking me for DNA information. And I thought this is just charred ruins."[74] Tom Rogér noted that many "airline families" did not think of the site as the final resting place of their loved ones.[75] He also wondered whether some of the families whose loved ones worked in the towers would even want to visit such a horrible place in the future.

In a moment of levity, Anthoula Katsimatides, who had worked for George Pataki and later led the LMDC's family relations effort, asked the jury to choose a design that had some color to it—she did not want to visit a memorial that was just gray steel—"can we throw in some red in there and . . . hot pink? Because I think that I want it to be a place that can make me smile. And I think my brother would have wanted me to be able to pray and cry, but [also] smile in his memory."[76]

Some family members mentioned the Libeskind plan and their concerns that it was being whittled away by the LMDC. Jack Lynch even told the jurors that he was concerned that the LMDC would overrule them in the interest of other powerful stakeholders. Kathleen Martens reiterated these concerns and told the jurors that they needed to think carefully about the integration of the memorial into the overall site plan—something that the Family Advisory Committee had been worried about since the redevelopment process got underway in mid-2002.

Other Stakeholder Meetings

As the meeting drew to a close, a few family members asked to meet again with the jury later in the process. Anita Contini informed participants of two additional chances to speak to the jury members at upcoming forums around the New York area. The next opportunity for stakeholders to address jury members came on June 5 at a meeting at Pace University between the jury and representatives of all 9/11 LMDC advisory councils plus representatives of the families of UA 93. This meeting covered little new ground, but offered all major stakeholder groups the chance to explain their views to the jury. Most notably, Edie Lutnick laid out the view that the needs of the families of victims ought to be paramount in the development of a memorial. Residents of lower Manhattan, including Marc Ameruso, disagreed with this claim. He reminded attendees that "there are a number of homeowners and home renters that live down there [here at the meeting]. The operative word is 'home.' This is our home. So to put more credence on one group than the other I think it's kind of disrespectful. . . . This is our home."[77] Local resident and Community Board #1 member Michael Connelly made a similar point, noting that "the people who live in Lower Manhattan are families also. My children were born here and have spent their entire lives in the shadow of the World Trade Center."[78] He went on to state that he wanted the site to be reborn as a vibrant neighborhood and not as a cemetery. Rather than invoking sites that had been preserved as memorials, such as Gettysburg and Pearl Harbor, Connelly invoked cities like London and Dresden that were nearly destroyed in wartime fighting but were rebuilt so that life could go on.

In addition to trying to shift the rhetorical focus of the site from a graveyard to "home," Ameruso also challenged the notion of "sacred ground."[79] He noted that he spent three-and-a-half days down at Ground Zero as a first responder and found remains not just in the sixteen-acre site but along West Street as well. He asked families to explain why they weren't advocating for the closure of West Street. Anthony Gardner continued this explicit effort to define terms: "To this gentleman over here who has redefined residents, I'd like to redefine September 11th families because we are a community. There's a September 11th community."[80] He believed that this community consists of survivors, rescue workers, families, and residents of lower Manhattan who do not share the view that the convenience of local residents ought to trump a dignified memorial to the victims of the attacks. He continued, "This is a sacred place. And to define sacred ground for you, sir, the area that we are specifically fighting for is to protect and preserve the area within the slurry wall known as the bathtub that comprises the Tower 1, Tower 2 footprints to bedrock, the Plaza area, World Trade Center 6 and the Marriot Hotel."[81] He (and later Lee Ielpi) agreed with Ameruso that remains were recovered far outside this area, but they were compromising and only trying to protect the area where the majority of remains were recovered.

Toward the end of his commentary, Gardner said that he was saddened that the press had been unfair to the families, by reporting that they were driven by emotion and unwilling to compromise. He argued that most families had been ready to make concessions from early on in the process—asking for the preservation of the bathtub area in exchange for the acceptance of commercial development throughout the rest of the site. Their main goal, he insisted, was to preserve the site in a meaningful way for future generations—not to compromise its integrity for short-term needs or the convenience of people around the site. Their vision was long-term: "This is America's memorial. This isn't only New York's. This isn't the September 11 families' memorial. No matter what, that will not bring my brother back. The only solace it will give me is knowing that history was preserved and future generations will learn from these martyrs, these people that gave their lives. . . . Look at Gettysburg. Look at Pearl Harbor. And that's what this site is."[82]

The LMDC also held an open public forum for all New Yorkers at the Tribeca Performing Arts Center at the Borough of Manhattan Community College on May 28, 2003. The discussion was similar to that at the first two events, with one new area of contention: whether rescue personnel should be given special treatment at the memorial. When Ric Bell, of New York New Visions, spoke in favor of a memorial that did not distinguish between rescuers and civilians, he was booed by the more than 150 firefighters who were there to protest the LMDC's decision not to give preferential attention to any group of names.[83]

The move to list rescue workers separately at the memorial was spearheaded by John Finucane, a retired firefighter who founded the group Advocates for a 9/11 Fallen Heroes Memorial. This group's mission was to ensure that "the 343 firefighters, 37 police officers of the PAPD and the 23 police officers of the NYPD, and all other uniformed rescue workers killed on 9-11, during rescue operations at the World Trade Center, be appropriately remembered, as part of the greater memorial. We ask that their names, along with rank, badge number and unit, be listed alphabetically, by department, just as they would want it. And that the memorial includes language that clearly acknowledges their sacrifice."[84]

Those who supported the Advocates' position offered justifications for differential treatment at this final forum. First, they argued that there was a fundamental difference between the civilians and the rescue workers who died on 9/11; the former group *lost* their lives, they believed, whereas the latter group *gave* their lives. Chris Ganci, whose father, Chief of Department Pete Ganci, died in the attacks, told the jury, "If it wasn't for them, you'd need twice as much space for a memorial, because there would be twice as many victims."[85] Second, they spoke of the unique bond among firefighters, police officers, and EMS workers who served as a unit—they went into the towers together, engaged in heroic behavior together, and died together. Thus, they deserved to be remembered together.[86] Finally, many Advocates spoke of the importance of their jobs for most rescue workers—it wasn't just what they did, it was who they were. The least the memorial could do was acknowledge this unique identity.

The placement of names would become a central focus of disputes over the memorial design—with rescue worker families and advocates, as well as many civilian family groups, making strong arguments against random placement and for the grouping of victims at the memorial.[87] But such conversations were still months away when the LMDC's public outreach campaign to inform the jury ended and memorial proposals began to arrive in advance of the deadline, set for June 30.

Eight Finalists

By 5:00 pm on Friday, 5,201 people had submitted their ideas for the memorial, making it the largest architectural competition of all time.[88] After a month of screening by LMDC staff, the jury began the arduous task of reviewing thousands of entries to select those few that were worthy of continued development in the second stage of the competition. According to James Young, jury members worked long days throughout August, examining approximately 600 designs each day, in a building overlooking the site. By the beginning of September, they had narrowed the field of 5,201 to 250, and after a further week of intense deliberations, narrowed that down to around twenty. At this point, in addition to deciding the merits of each design, the jurors had a lengthy and contentious debate about whether to publically reveal the finalists before selecting a winner. According to Young, the architects on the jury worried that a public display of finalists would subject them to intense lobbying from friends and colleagues. He and others, however, argued that the integrity of the process—the LMDC had already promised a public showing—and the legitimacy of the final decision required that the public be allowed to see and comment on the finalists before a winning design was chosen.[89] Plus, there was always the possibility that the public would notice something, positive or negative, about the designs that the jury had missed.

By mid-October, the jury had come to two realizations: first, they could narrow the list of finalists to eight but no fewer, given that no clear favorites had emerged.[90] And second, they were going to be an "interventionist" jury, in the sense that they were going to request changes to each of the final designs before selecting a winner. According to Young, "we could not recommend any of the designs without significant modifications, what amounted to a kind of 'critique' architecture students receive during their training. The LMDC approached each of these teams, still unidentified to the jury, and provided them with our detailed critiques and funds for developing their boards into three-dimensional models and animating their designs with a short computer-generated walk-through."[91] Once the eight finalists completed this work, they met with the jury in person in mid-November to discuss their visions in detail.

A few days later, on November 19, 2003, the finalists were introduced to the public and their designs were put on display in the Winter Garden of the World Financial Center.[92] They were:

- "Votives in Suspension" by Norman Lee and Michael Lewis, which featured subterranean sanctuaries in each footprint with hanging lights, each representing a victim, and green space in the depressed area set aside by Libeskind for the memorial.
- "Lower Waters" by Bradley Campbell and Matthias Neumann, a multilevel memorial museum that featured water elements and landscape and hewed largely to Libeskind's master plan for the site.
- "Passages of Light: The Memorial Cloud" by Gisela Baurmann, Sawad Brooks, and Jonas Coersmeier, in which a translucent white cloud-like structure made of plastic and glass seemed to hang over the bathtub area and represent a bandage over a wound.
- "Suspending Memory" by Joseph Karadin with Hsin Yi Wu, which imagined the tower footprints as island gardens surrounded by water and connected by a bridge.
- "Garden of Lights" by Pierre David with Sean Corriel and Jessica Kmetovic, which combined garden elements above ground with underground rooms in which spotlights represented victims.
- "Reflecting Absence: A Memorial at the World Trade Center Site" by Michael Arad, which combined a stark plaza at street level with voids where the footprints of the towers were, and an underground sanctuary to remember and mourn the dead.
- "Dual Memory" by Brian Strawn and Karla Sierralta, which included an above-ground landscape of trees and an underground chamber with light portals that glowed with a unique intensity (each one representing the individuality of a victim) along with images of victims projected on glass panels.
- "Inversion of Light" by Toshio Sasaki, which in Goldberger's words "called for the footprint of the south tower to become a reflecting pool above a circle of lights. The floor plan of the north tower would be illuminated from below and a blue laser light would shine into the sky."[93]

In their official statement, the jury noted that "the eight memorial designs chosen as finalists have a number of characteristics in common. They strive neither to overwhelm the visitor nor their immediate surroundings. They aspire to soar—not by competing with the soaring skyline of New York but rather by creating spaces that strive to reconcile vertical and horizontal, green and concrete, contemplation and inspiration. They allow for the change of seasons, passage of years, and evolution over time. They emphasize the process of memorialization over their own grandeur and present themselves as living landscapes of living memory that both connect us to our past and carry us forward into the decades ahead."[94]

The jury then noted that none of the designs was perfect and that all were "still in development." Each design would require "additional refinements, including how the names of the victims should be recognized, how to respect the tower footprints and keep them unencumbered, how to provide access to bedrock, and what the relationship of the memorial will be to the site's interpretive museum." The jury concluded with the statement that "the memorial itself is a process, an attempt to bring reconciliation to that which can never be reconciled: love and loss, heroism and horror, past and present, public recognition and private introspection. The power and pain of this memorial's public discussions energize and animate this process, and thus help keep this memory alive."[95] The translation seemed to be: none of these eight proposals are perfect, we are going to do everything in our power to make them as good as they can be, and the public should expect there to be significant debate about the merits of each design and whether any of them adequately do the job they are supposed to do.

Public Response to the Finalists

Initial public reaction to the eight finalists was decidedly mixed, and media coverage highlighted this ambivalence. Many commentators, whether they were architecture fans or families of the victims, decried the designs as impersonal and overly abstract. They were generic modern structures that said nothing about what happened at the World Trade Center and could just as easily have doubled as a hotel lobbies, airport concourses, or restaurants. Others praised the designs, arguing that their abstract nature left room for contemplation and personal interpretation. Proponents also believed these designs had surpassed the classical notion of building a static monument in favor of active memorialization of the victims. Nevertheless, *New York Times* architecture critic Herbert Muschamp spoke for many when he criticized all eight designs for over-embellishment. In his view, they did not live up to the standards set by juror Maya Lin's Vietnam Veterans' Memorial. Looking for something positive to say, Muschamp singled out Michael Arad's Reflecting Absence as the one that "has the greatest potential to be the least."[96]

Although none of the designs actually hewed to Libeskind's site plan, most did build down into the ground rather than above it. Many firefighter families were irate that none of the proposed memorials separately honored the rescue workers who died in the call of duty, and the Coalition of 9/11 Families was deeply disturbed that none of the proposed designs preserved the footprints in a historically accurate and meaningful way. Each design recognized the footprints, but none actually preserved them in the plans made public on November 19. In a press conference held in a hotel near the Winter Garden, the Coalition of 9/11 Families issued a report card in which all eight designs received an "F" for their failure to preserve the footprints, whatever their other merits.[97]

Just as striking as what the final designs contained was what they did not: no representations of, or even allusions to, the former twin towers; no representations of what actually happened on September 11, whether realist or symbolic; no patriotic symbols such as the American flag; and no architectural remnants from the fallen towers. There were certainly such designs among the submitted entries, but the jury clearly favored abstract proposals for the memorial.

Although there was no formal mechanism for the LMDC to gather public comments on the eight designs, Imagine New York, a project of the Municipal Arts Society, decided to hold "Toward the People's Memorial" in the days after the designs were released.[98] The goal of this event was to ensure that ordinary citizens could voice their opinions about the memorial proposals. Over the course of six sessions, more than 300 people participated in workshops held in the New York area and more than 12,800 people registered a total of 15,000 comments on the website Imagine New York set up as an online forum.

As with the jurors, no clear favorite emerged among participants. The general consensus seemed to be that, although all eight designs were passable, none fully met public's expectations. Participants liked elements of each, and some suggested that a mix-and-match approach might be the best way to move forward. Many participants also felt that all eight finalists catered more to the needs of victims' families than other stakeholders. There was a sense that the designs chosen were lifeless, cold, and focused on loss, with little acknowledgement of the vibrancy, diversity, community spirit, and continuous renewal present in the city. Many participants, echoing Goldberger's criticism, also decried the blandness and lack of site-specificity. The eight designs said little about the World Trade Center site or what happened there on September 11. Participants also complained that the finalists tried too hard to make something artistic, architectural, and beautiful out of an ugly, horrible, and cruel event. While nobody suggested graphic depictions of death and violence, they did want to see some allusion to what happened at the site—for example some of the damaged steel girders or Koenig's *Sphere*. Finally, many participants expressed confusion about the relationship between the memorial and the museum, and others were worried about long-term maintenance issues associated with particular designs. There was also discussion of the placement of names on the memorial, but no consensus was reached about the best approach.[99] Similar sentiments could be found on Internet bulletin boards such as Wired New York, in the comment sections of major media outlets, and on blogs that regularly focused on the redevelopment of the World Trade Center site.

In its letter to the competition jury, New York New Visions summarized the comments it had received from the public and offered its own, more pessimistic view of the finalists: "We feel . . . that the eight designs taken as a whole represent only one narrow interpretation of the guideline" and ultimately "suffer from sameness. They are all from one family of design approaches."[100] In the end, New York New Visions suggested that the jury return to the designs it had previously rejected to look for something new and fresh or work with Daniel Libeskind to further refine and develop what the jury considered to be the best of the eight designs. New York New Visions also asked the jury to recognize that site planning and the memorial competition could no longer proceed in isolation—this bifurcated process was producing less than desirable results. It was, in essence, a statement from New York New Visions that, so far, the process had failed.[101]

New York New Visions was not alone in this sentiment. Governor Pataki let it be known that he was not impressed with the finalists.[102] Former Mayor Rudolph Giuliani felt the same way.[103] So did *New York Times* art critic, Michael Kimmelman, who argued that the LMDC ought to scrap the open competition and commission a great work of art by a leading designer. This was, he argued, how St. Louis got its iconic arch.[104] Except that it wasn't: Eero Saarinen's arch was not the product of a commission, it was the product of the same kind of competition sponsored by the LMDC. This error, according to Goldberger, made the jury even more determined to produce a fitting and beautiful memorial at the World Trade Center.[105]

For the jurors, all of this public conversation was initially a cause for concern because they feared that consensus would emerge about a single design before they had a chance to come to their own decision. This would then put them at risk of choosing a design that was not favored by the public. As it turned out, the jury had nothing to worry about—while many designs had a significant public following, "none garnered anything approaching public acclamation."[106]

A Winner Emerges

While the public discussed and debated the merits of each proposal, the jury was "struggl[ing] desperately to whittle the number of finalists down to three by the end of the year, and finally did so only with great difficulty."[107] Even when they did get to three—Garden of Lights, Passages of Light: The Memorial Cloud, and Reflecting Absence—Young said that "it was not clear whether we would be able to rally around only one with anything approaching the fervor and excitement we were all looking for."[108] In an effort to generate a clear winner, the jury asked for even more extensive changes to these three designs.

Of the three finalists, Maya Lin was widely reported to be partial to Michael Arad's Reflecting Absence, but several jurors were concerned that his stark plaza with two box-shaped voids would seem cold and off-putting when built. They thought that adding landscaping would make it more comforting and inviting. At the behest of the jury, Arad formed a partnership with Peter Walker, a well-known Berkeley-based landscape architect with a taste for minimalist design, who had submitted his own memorial design that hadn't made the final cut. Walker accepted the job and immediately set about augmenting Arad's austere plaza with a forest of trees and landscaping.[109] In addition to enhancing the design of the street-level plaza (which was a rejection of the sunken memorial space that Libeskind proposed), Arad also added access to the footprints and slurry wall, as well as a space for a memorial center, below ground.

As the final deliberations got underway in December, Gardens of Light was losing support, both because the designers didn't make the changes requested by the jury and because the notion of turning the World Trade Center, at the heart of one of the most dense urban neighborhoods in the world, into two box-shaped wildflower meadows and a fruit orchard ultimately did not make sense. Besides upkeep issues, the jury was concerned that such a memorial would not adequately represent the horror that took place on September 11—it would be all redemption but no loss and pain.[110]

With Garden of Lights out of the running, only Arad's Reflecting Absence and Passages of Light: The Memorial Cloud by Gisela Baurmann, Sawad Brooks, and Jonas Coersmeier were still in the running. The jury was split roughly evenly between each design, with Enrique Norten heavily favoring The Memorial Cloud, and Lin, Van Valkenburgh, and Young being partial to Reflecting Absence. Of the two, The Memorial Cloud was more visually stunning, with its crystalline glass and translucent plastic cloud seemingly floating above the bathtub area. Despite its beauty, many jurors were concerned that it lacked an obvious memorial function. In other words, besides the reference to the dust cloud that enveloped lower Manhattan in the moments after the towers collapsed, the design did not stir the memory in any essential way.[111]

In early January 2004, after weeks of deliberations, the jury finally voted in favor of Arad's design. According to Young, the decision boiled down to the fact that Reflecting Absence captured the "twin motifs of loss and renewal" better than Memorial Cloud.[112] At the heart of Arad and Walker's design were two large voids containing recessed pools and waterfall elements that sat within the footprints of the towers. These voids, which the designers described as "open and visible reminders of the absence," were located in a field of trees that would serve to integrate the memorial into the city's fabric and the daily lives of New Yorkers. In the initial design, a pair of ramps that led into the memorial bordered each pool. The function of these ramps was to transition visitors "from the sights and sounds of the city," into the contemplative memorial space. Describing the landing at the bottom of the ramps, they explained that visitors would "find themselves behind a thin curtain of water, staring out at an enormous pool. Surrounding this pool is a continuous ribbon of names. The enormity of this space and the multitude of names that form this endless ribbon underscore the vast scope of the destruction. Standing there at the water's edge, looking at a pool of water that is flowing away into an abyss, a visitor to the site can sense that what is beyond this curtain of water and ribbon of names is inaccessible."[113]

Reflecting Absence also contained numerous underground spaces for contemplation and memorial services, areas where the slurry wall and original foundations could be viewed, and an underground interpretive center and research library located at bedrock. The design included a remains repository that would be situated in a room seven stories below ground at bedrock of the North Tower footprint. According to Arad, "here a large stone vessel forms a centerpiece for the unidentified remains. A large opening in the ceiling connects this space to the sky above, and the sound of water shelters the space from the city. Family members can gather here for moments of private contemplation. It is a personal space for remembrance."[114]

The choice of Reflecting Absence transformed thirty-four-year-old, Israeli-born Michael Arad from an unknown architect for the New York City Housing Authority into the designer of one of the most important structures of the early twenty-first century. As *New York Magazine* put it, "he looked like the Maya Lin of 9/11—a bright, shining star out of nowhere who would build a breathtaking new landmark."[115] Unfortunately, things would not work out as well for Arad as they did for Lin. "Two years later," the 2006 article continued, "all we've got is a pile of dirt, a price estimated nearing $1 billion, and a nasty behind-the-scenes war of wills."[116]

6

Remaking the Memorial

Just as LMDC was introducing Reflecting Absence, and Michael Arad, to the public, another aspect of Daniel Libeskind's plan to redevelop the World Trade Center site—the single skyscraper meant to replace the twin towers in the skyline and in the mind's eye—was devolving into what Elizabeth Greenspan describes as the "white elephant" of the rebuilding effort.[1] Libeskind had conceptualized World Trade Center 1 as a symbol of American democracy and renewal, complete with "sky gardens," a height of 1,776 feet, and a form and location that would echo the profile of the Statue of Liberty. Governor Pataki bought into Libeskind's vision and described the skyscraper as "not just a building. This is a symbol of New York. This is a symbol of America. This is a symbol of Freedom."[2]

Developer Larry Silverstein, on the other hand, was interested in a building that would attract renters and produce profits. From the beginning, he made it clear that he was going to play by his own rules. In a move demonstrating his power over the site, he forced Daniel Libeskind to partner with his favored architect, David Childs, from the historic architectural firm Skidmore, Owings & Merrill, to redesign the tower to his liking. The resulting fight produced a design that was quite different from what Libeskind had originally proposed. Add to this law enforcement concerns over the safety of the building that would at once be a symbol of resilience and a potential target of future terrorist attacks, and their demand that the base of the tower be fortified with concrete, and it appeared to the public that nothing could happen at Ground Zero without an unseemly collision of business, politics, and raw emotion.[3]

Ultimately, in 2006, Silverstein and the Port Authority would negotiate a deal that turned over the Freedom Tower (which would eventually be renamed "One World Trade Center") to the Port Authority in exchange for $3.5 billion, a reduction of the rent payments Silverstein was required to make on the site, and a guarantee that the Port Authority would rent a significant amount of office space in the properties Silverstein developed at prevailing market rates. Although the deal was an odd one from the perspective of the Port Authority, especially given the fact that it had signed the lease of the World Trade Center over to Silverstein precisely to get out of business of property management, it at least ensured that construction of the new office building would eventually begin. It also demonstrated the importance of the Freedom Tower concept to Governor Pataki, who wanted to make sure that the project was far enough along before he left office that the next governor couldn't pull the plug.[4]

Making the Memorial

In the beginning, there was hope that the memorial design process would be more uplifting and less dysfunctional than the Freedom Tower debacle. The youthful Michael Arad came across as genuine and earnest, even to those 9/11 family members who disliked his design. But this public persona belied a fierce independence and stubbornness that gave him the courage to ignore much of Libeskind's master plan. Arad abandoned the notion of a sunken memorial (except, of course, for the waterfalls) and eliminated the cultural building that Libeskind designed to cantilever over the North Tower footprint. He even changed the access plan to the site developed by the Port Authority.[5] Libeskind, not surprisingly, was unhappy when he heard that Arad's design was chosen, but quickly decided that it would be better for him to work with Arad and Walker to make the design fit than to advocate for its outright rejection. He recognized that the jury process could not be undone, and he was stuck with its decision.[6]

LMDC officials expressed excitement about Arad and his design when they were introduced to the public on January 14, 2004. Behind the scenes, however, an intense effort was undertaken to keep Arad in his place and to harmonize Arad's design with Libeskind's master plan and the many requirements of the site. Arad, Walker, Libeskind, and LMDC made a series of compromises in tense talks that took place between the parties in early January 2004.[7]

In addition, because of the technical complexity of the situation (especially the transportation and facilities infrastructure that needed to be fit into the memorial site), LMDC brought in Max Bond from the firm Davis Brody Bond as an "associate architect" on the project to help "realize" it.[8] Bond was a major figure in New York architecture with experience designing major projects for the Port Authority and other top-tier institutions, including the Lincoln Center and the New York Public Library. He was also already familiar with the site from his role as a member of LMDC's Memorial Center Advisory Committee. In addition, his firm had been involved in the creation of several important memorials around the country, including the Civil Rights Institute in Birmingham, Alabama and the Martin Luther King, Jr. Center in Atlanta.[9] Arad, however, resisted the decision to force him into such a partnership, because he knew that an architect like Bond would be able to exert tremendous a control over the design refinement process.

The clash of personalities among Arad, Walker, Bond, and Libeskind has been well documented.[10] Their struggles were many, but the bottom line was that Arad resented being forced to compromise his vision, Libeskind resented Arad for blithely ignored his master plan, and Walker and Bond resented Arad for being unwilling to heed their advice and recognize that it was futile to battle the Port Authority. Indeed, Bond's firm had a reputation for being practical and putting the client's satisfaction ahead of obedience to pure artistry.[11] LMDC officials did their best to mediate among the parties and all sides recognized they needed to find common ground, even if the process was highly acrimonious. But the mediation failed, and the project stalled for months.

Meanwhile, a controversy emerged over the LMDC's desire to make the 9/11 memorial quadrant a celebration of life, and not just a remembrance of death and destruction. They sought to balance the memorial and museum with four cultural institutions that, while invoking the spirit of 9/11, would transcend it to promote enhanced understandings of the broader world and showcase lower Manhattan as a lively, vibrant 24/7 neighborhood. While residents largely applauded this decision, families and many rescue workers felt that promoting culture on the World Trade Center site was an inappropriate way to memorialize those who had died there.

The International Freedom Center

In June 2003, the LMDC invited cultural institutions to submit proposals to become a part of the World Trade Center community; it received 113 applications by September. In June 2004, the LMDC announced the winners: the Joyce International Dance Center and the Signature Theatre Company would occupy a performing arts center, while the Drawing Center (an institution devoted solely to the curation and display of drawings) and a newly created "International Freedom Center" would be housed in a museum complex.[12]

Many 9/11 families did not want space at the memorial site to be used for any purpose other than telling the story of 9/11 and its victims. Although these families objected to the placement of arts institutions at the site, they reserved their strongest ire for the International Freedom Center (IFC). The IFC was developed specifically for the World Trade Center site by three men with an intense interest in promoting the spread of liberty and freedom around the world in the post-9/11 era. The first, Tom Bernstein, was a wealthy lawyer and investor who had made a fortune in the entertainment industry and in developing Chelsea Piers. He was also the president of the board of the legal advocacy group Human Rights First, the organization that filed lawsuits against the U.S. government on behalf of accused terrorists in Iraq and Afghanistan and that has been central in the fight to close Guantanamo. Bernstein was joined by Peter W. Kunhardt, whose company produced documentaries for PBS and other outlets, and Richard Tofel, a former assistant publisher at the *Wall Street Journal* and vice president of Dow Jones & Company. The IFC's board included Bernstein; FAC member and memorial competition juror Paula Grant Berry; Stephen J. Heintz, President of the Rockefeller Brothers Fund; and Daniel T. Tishman, President of Tishman Construction Corporation. It was advised by a who's who of well-known academics, journalists, and civil libertarians.[13]

The mission and vision of the IFC was to link the story of 9/11 with the broader struggle for freedom over time and to encourage visitors to fight for freedom wherever it was not yet attained. The IFC's mission, global in scope, was to illuminate "humankind's sometimes uneven but ultimately enduring aspiration for free and open societies."[14]

Families who lost loved ones on 9/11 believed the IFC mission would detract from the story of what happened that day. They also worried that the IFC would seek to explain the actions of the terrorists, even though they were, in the eyes of the families, senseless and ultimately inexplicable. The memorial should offer visitors a factual recounting of what happened on 9/11 (and in 1993), the stories of the innocent victims whose lives were cut short, and nothing more.

In a *Wall Street Journal* opinion piece that would ignite the anti-IFC movement, Debra Burlingame, a member of the World Trade Center Foundation Board whose brother was Charles F. "Chic" Burlingame III, the pilot of the flight that crashed into the Pentagon on September 11, 2001, wrote that visitors to the 9/11 memorial "will expect to see the artifacts that bring [their] memories [of 9/11] to life. They'll want a vantage point that allows them to take in the sheer scope of the destruction, to see the footage and the photographs and hear the personal stories of unbearable heartbreak and unimaginable courage. They will want the memorial to take them back to who they were on that brutal September morning."[15] Instead, because of the IFC, she wrote, they would get an ideologically biased history lesson and lectures on the meaning of liberty in the post-9/11 world. Rather than learning about 9/11, they would learn about the effect that Abu Ghraib will have on global geopolitical stability. Such events were not directly relevant to the planes crashing into the twin towers, she argued, and would crowd out the artifacts and stories that could tell us about what, exactly, happened on 9/11. "The public will have come to see 9/11," she wrote, "but will be given a high-tech, multimedia tutorial about man's inhumanity to man, from Native American genocide to the lynchings and cross-burnings of the Jim Crow South, from the Third Reich's Final Solution to the Soviet gulags and beyond. This is a history all should know and learn, but dispensing it over the ashes of Ground Zero is like creating a Museum of Tolerance over the sunken graves of the USS *Arizona*. The public will be confused at first, and then feel hoodwinked and betrayed. Where, they will ask, do we go to see the September 11 Memorial?"[16]

Inspired by Burlingame's *Wall Street Journal* editorial, a conservative blogger named Robert Shurbet founded the group Take Back the Memorial to advocate for "a fitting and proper memorial to be built for those who perished on September 11th, and to tell the story of that fateful day—and that day alone" in a way that was "free of politics."[17]

This organization, while it began primarily as an Internet forum, quickly gained the support of a group of powerful 9/11 family group activists (many of whom were involved in the Coalition of 9/11 Families), including Monica Iken, Lee Ielpi, Sally Regenhard, Charles Wolf, and Michael Kuo. In addition to their collective belief that an institution like the IFC was inappropriate for the World Trade Center site, many relatives worried that the above-ground glass building designed by the world-class Norwegian architecture firm Snohetta would overshadow the actual memorial, which was slated to be underground. They also believed that the IFC and the 9/11 memorial would compete for fundraising dollars. Edie Lutnick signed on to the anti-IFC efforts for this reason. In a speech on June 20 organized by Take Back the Memorial and fifteen other family groups, Lutnick told the audience that visitors to the World Trade Center should be immersed in 9/11 and not world politics. It should be a place of sacred remembrance and not a speaker's corner. "Make no mistake," she concluded, "we think Martin Luther King deserves to be honored, that there is a place for Ukrainian [victims] to be studied, but not on the site where 20,000 body parts of 2,749 innocents were recovered."[18]

What was different about Burlingame, compared to the families that comprised the coalition, was her willingness to turn what had formerly been a fight over urban redevelopment and historical memory into part of the broader culture wars. Despite the fact that the people involved in the IFC were a decidedly bipartisan group (Bernstein was a close friend and former business partner of George W. Bush, after all, and the board also included John Raisian, director of the Hoover Institution, a conservative think tank at Stanford University), Burlingame singled out the most liberal-leaning advisors and supporters, including historian Eric Foner, ACLU executive director Anthony Romero, Judge Michael Posner, and financier and philanthropist George Soros. "The so-called lessons of September 11 should not be force-fed by ideologues hoping to use the memorial site as nothing more than a powerful visual aid to promote their agenda. Instead of exhibits and symposiums about Internationalism and Global Policy we should hear the story of the courageous young firefighter whose body, cut in half, was found with his legs entwined around the body of a woman. Recovery personnel concluded that because of their positions, the young firefighter was carrying her," she wrote.[19] For the families of the coalition, who had formerly relied on their moral standing, invoking the culture wars was certainly a new strategy.[20] It was one they adopted, however, in an effort to reserve the memorial quadrant for the memories of their loved ones. Indeed, it seems that the final line of Burlingame's editorial clearly resonated with them: "Ground Zero has been stolen, right from under our noses. How do we get it back?"

Take Back the Memorial quickly mobilized protests against the IFC, threatened to derail fundraising for the memorial, and waged a highly effective media campaign to convince the public that the IFC (and the Drawing Center, too) would promote a "blame America" agenda.[21] In response to the uproar, on June 24, Governor Pataki, who had until then been a vocal supporter of the IFC and the Drawing Center, asked the institutions to promise to respect the sanctity of the site and not include any content that denigrated America, the response of local and national authorities to the attacks of 9/11, or the notion of freedom itself.[22] As journalist Robert Kolker noted, this request "was the first step in a long, slow public hanging," that would end on September 28, 2005 when Governor Pataki officially pulled the plug on the IFC.[23] Supporters of the project simply could not overcome the backlash created by Take Back the Memorial.

Put It Above Ground

The successful Take Back the Memorial campaign seemed to embolden the core group of 9/11 family activists. In March 2006, another group emerged to fight what it perceived as serious problems with the Reflecting Absence design, especially the placement of the names of the dead underground, and fears that visitors to the memorial would become trapped underground in the event of an emergency or terrorist attack (a much-discussed scenario was a bomb being thrown down one of the voids into the underground chamber of the memorial) or fire. The core argument of this group was that it was both disrespectful to the dead, and unsafe for the living, to have an underground memorial.[24] In their view, the desire of families to preserve the footprints of the twin towers was used as an excuse to hide the memorial to the dead from view and to sanitize the site at street level for the benefit of residents and businesses. The only acceptable solution was instantiated in the name that they chose for their organization, "Put It Above Ground," and their motto, "never forget, safety first." In addition to an aboveground memorial, the organization also advocated for the preservation of a piece of the façade of the twin towers to be displayed at Ground Zero to remind visitors of exactly what happened that day, as well as the preservation of the footprints at bedrock so long as a safe emergency exit plan could be developed. They also rejected Arad's plan to randomly place the victims' names on the memorial.[25]

Put It Above Ground was primarily a collaboration between three 9/11 activists—Rosaleen Tallon, Sally Regenhard, and Dennis McKeon—as well as many other family members and individuals who supported the cause. Tallon, who lost her brother, firefighter Sean Tallon, in the attacks, was an active member of Advocates for a 9-11 Fallen Heroes Memorial, the organization pushing for special recognition of uniformed service personnel at the memorial. Through her advocacy work, she met Regenhard in 2005. She was shocked to learn that the proposed underground memorial design lacked what Regenhard considered safe exits, and seemed to invite acts of terror with its open roof. More generally, Tallon was disturbed to learn from Regenhard that the Port Authority did not need to comply with New York City building codes. Although he didn't lose any family members in the attacks, McKeon quickly became a central disseminator of information about services available in the borough (and around the region) to those affected by the attacks. His Internet bulletin board, where-to-turn.org, quickly became the place where families and ordinary people concerned about the redevelopment of the World Trade Center site went for information, and it remains a tremendously valuable repository of information about 9/11 activism today.

In an effort to get the attention of the LMDC, Put It Above Ground members staged a rally near the site on February 27, 2006, asking the LMDC to address their concerns about the underground memorial. LMDC's response was essentially to ignore its substance. LMDC president Stefan Pryor wrote to Tallon in the days after the rally: "Again, thank you for your concern and passion. We share the goal of creating an appropriate and magnificent memorial at the World Trade Center and that is what we will continue working on."[26]

Upset but not surprised at being blown off by LMDC, Tallon decided to do something dramatic. First, she started an online petition asking people to support their campaign to get the LMDC to build an above-ground memorial with names placed in logical groups and a design that met New York City safety codes. According to McKeon, more than 14,000 people ultimately signed this petition.[27] Second, and more dramatically, she slept in front of her brother's firehouse (Ladder Company 10), directly across from the World Trade Center site, for seventeen nights in March 2006.[28] This vigil received a great deal of media attention, and made concerns about the placement of names below ground, and disputes about the memorial, a public matter.[29]

Not all family members agreed with the stand taken by Put It Above Ground. Indeed, both Monica Iken, who had been most dedicated to creating a memorial for the victims of 9/11, and Mary Fetchet, a founding member of the Coalition of 9/11 Families, publically supported Arad's underground design. In their view, it had both symbolic value, in the sense that it allowed family members to descend toward bedrock when they visited the site, and it shielded the memorial from the noise and chaos of the city.[30] The LMDC would eventually be forced to address these issues, along with the soaring projected cost of the memorial, whether it wanted to or not. Tallon's actions and subsequent protests and rallies would also make Put It Above Ground a force to be reckoned with in future debates about the development of the memorial and museum—particularly those concerning human remains.

The Repository

Another controversial aspect of the below-ground memorial was how to store the unidentified and unclaimed remains of victims of the World Trade Center attacks held by the OCME. Families only became aware of the details of Arad's plan when the first set of construction documents were released by the LMDC to get bids on pouring the Memorial's concrete footers.[31] According to construction documents, the North Tower footprint was to include a central 100- by 100-foot "contemplation room" with a nine-foot-high, black stone "symbolic mortuary vessel" that would represent the victims of the attacks, but would not actually contain their remains.[32] The thousands of remains would be stored in a climate-controlled 53- by 34-foot storage chamber behind a wall of the contemplation room. This chamber would not be publically accessible and would be controlled by the OCME. OCME officials would be able to enter the storage area and remove remains for further testing or for handover to families when appropriate.[33]

According to preliminary plans, visitors to the site would have been able to access this contemplation room through a corridor from the section of slurry wall that was being preserved to bedrock, while families would be able to go straight to the contemplation room by private elevator. The North Tower footprint would have also contained a 2,400-square-foot family room that was open only to relatives of the victims, as well as a much smaller 360-square-foot room adjacent to the actual storage facility that would allow families to look into the chamber to see how the remains were being kept. The design published by the LMDC also included two exhibition rooms around the contemplation room that could be used for displays about the victims or the events of 9/11, as well as significant walled-off space for mechanical infrastructure for the memorial and the waterfalls.[34]

The Coalition of 9/11 Families was deeply disturbed when they saw this design in the *New York Times* on January 3, 2006.[35] They objected both to the inclusion of a "symbolic mortuary vessel" and to breaking up the North Tower footprint into several rooms. In their view, the actual box columns and the original concrete slab at the base of the towers would still resonate with visitors in thousands of years—in the way ruins of ancient structures continue to draw us today. These would give visitors an authentic, tangible historical connection to the site and those who died there, and would provide a sense of just how large each of the towers had been. To break this space into small rooms was tantamount to defiling the site's sacredness and historical authenticity. Further, most family groups argued that the repository should not be placed within either footprint, which would, in effect, privilege one set of victims over others—rather, it should be built somewhere within the slurry wall.[36]

More problematically, the notion of a "symbolic mortuary vessel" was, for many, deeply offensive. Several coalition members, as well as others, including Diane Horning and Edie Lutnick, noted that their loved ones were actually dead, not symbolically dead, and deserved an actual tomb where families could go to mourn their loss. Lutnick asked, "What is the purpose? The ashen remains are in Fresh Kills in a garbage dump. Now, on the site itself, it's going to be an empty box."[37]

Another issue that troubled many relatives was the placement of the family room several stories underground. They argued that this space should be above ground overlooking the memorial. At the time, the coalition and other family groups did not object to the notion that the repository itself should rest on bedrock, so long as it was designed properly. Indeed, when the LMDC's Family Advisory Council drafted their memorial program recommendations in 2003, they called for "a separate accessible space to serve as a final resting-place for the unidentified remains from the World Trade Center site," the exact language that was included in the memorial competition guidelines.[38] The Coalition of 9/11 Families believed that "the Memorial area must secure a repository in the area encased by the retaining walls for all the unidentified and unclaimed remains to be interred."[39] Further, the coalition firmly believed that sacred space extended from "bedrock to infinity" and wanted the memorial to be located at seventy feet below grade, not thirty feet as in Libeskind's master plan, which had been modified to accommodate transportation infrastructure.[40] Summarizing their position in 2003 (which appeared with slightly different wording in many places), the coalition wrote that "the Memorial Complex should be encased in the slurry wall from bedrock to surface level. . . . The Coalition supports returning all unidentified and unclaimed remains to the site in a place of reverence and honor surrounded by a place of reflection."[41]

In an effort to stop the LMDC from moving forward with the plan described in the bid package, the coalition staged a protest at the site and filed a petition against the corporation in the New York Supreme Court.[42] The coalition asked the court to halt all construction until the LMDC could engage in additional consultations about the preservation of the footprints at bedrock. The court ultimately rejected this request on the grounds that the LMDC had already engaged in extensive consultation with families and other stakeholders, that it entered into a historic preservation agreement in 2004, and that this agreement never included the actual concrete slab at the bottom of the bathtub as an element that was to be preserved. LMDC did, however, commit to preserving as many of the box columns as possible. The court also noted that the many stakeholders approved of LMDC's plans and had filed amicus curiae briefs against the coalition's petition. Finally, the court disputed the coalition's claim that the LMDC was required to consult with interested parties each time it made a change to its plans. In the court's view, such consultations were completely voluntary after the programmatic agreement (i.e., which elements of the structure were to be preserved) was reached. The court concluded by sympathizing "with the anguish that has fueled petitioners' quest for what they perceive as the ideal symbol of their loss," but ultimately found that the historical review requirements had been more than met by the LMDC.[43] Although this decision settled the matter legally, the storage of human remains at Ground Zero would come to be a major source of contention between the LMDC and their allies and many of the activist families from Put It Above Ground, Take Back the Memorial, and the Coalition of 9/11 Families when plans for the memorial were radically revised later in the year.

Ordering Names

In addition to debates over the structural and safety aspects of the memorial, the placement of names was also a major point of contention for many families. Although Arad wrestled with many different ways of arranging the names of victims around the pools of his below ground sanctuary, he ultimately decided that they ought to be arranged "in no particular order." To do it any other way would arbitrarily create a false order out of a chaotic event and would potentially upset families whose loved ones did not fit into any particular group. "The haphazard brutality of the attacks," Arad noted in the official description of his design, "is reflected in the arrangement of names, and no attempt is made to impose order upon this suffering." As far as special recognition for rescue workers was concerned, they could be "acknowledged with their agency's insignia next to their names." In order to locate the name of an individual, visitors could request help from on-site staff or a consult a printed directory.[44]

This solution to the names problem was welcomed by some families, but disliked by many others. Some civilian families did not like the idea that rescue workers would get special recognition on the memorial, while rescue worker families were upset that so little information would be printed about their loved one and that they would not be listed in a single place with their comrades. Some civilian groups that represented large constituencies of families were upset that victims who worked together or had other connections would not be found on the memorial in the same place.

In an effort to bridge all of these constituencies, Edie Lutnick, from the Cantor Fitzgerald Relief Fund, undertook a massive effort to bring all stakeholders to the table to come up with a solution that everybody could live with. While the LMDC cited these divergent opinions as evidence of infighting and discord, Lutnick believed that the LMDC had played various family groups against one another in order to do exactly what it wanted to do. At the first meeting of family group representatives, Lutnick realized that, although each had their own ideas about how the memorial would ideally represent their loved ones, with a great deal of work, common ground could be found. It took eight months to devise a solution that "a vast majority" of the thirty-two family groups involved would be willing to sign off on, which, Lutnick suggested, represented about two-thirds of the families.[45]

Lutnick described the final product of their deliberations as "straightforward. Everyone would be grouped and identified by their affiliation. Civilians would be listed in alphabetical order within their affiliation. Uniformed services workers would be listed and identified by rank. Each person who died would have their age and their floor following their name."[46] Further, the rescue workers' families wanted a separate place for their loved ones names because it was unclear in what tower many of them died. Because they would be listed by affiliation, they agreed that insignias were no longer necessary.[47] On October 8, 2004, Lutnick sent the compromise proposal to Governor Pataki, Mayor Bloomberg, and LMDC/WTC Memorial Foundation Chairman John Whitehead. In her account, she got only a cursory response from the LMDC stating that their proposal was one of many received. Over the next few months, Arad and his supporters remained steadfastly committed to a random placement of names.

Lutnick was motivated to join this effort because she felt strongly that the lack of historical context in Arad and Walker's design meant that the placement of names would have to tell the story. Further, she argued that there was nothing random about the well-orchestrated terrorist attacks that led to the murder of 3,000 people. "Listing the names arbitrarily," she writes, "would dehumanize people who died and deny the visiting public an emotional connection to them. . . . The 658 people at Cantor Fitzgerald had a connection to each other and their names should not be scattered as if they didn't."[48]

While the *New York Times* had offered families of the victims sympathetic coverage on numerous instances, in this case, the newspaper came out strongly against Lutnick's compromise proposal. In an editorial from July 5, 2006, the paper wrote that "we have always respected the emotional burden the 9/11 families have had to bear, as well as the complicated ways that private grief intersects with public issues during the rebuilding of Lower Manhattan. But when it comes to the heated debate over how the names of the victims of the World Trade Center attack are to be placed at the ground zero memorial, we are simply puzzled."[49] The editorial argued that the "natural analogy" for Arad's design was Maya Lin's Vietnam Veterans' Memorial, where the names are listed in the order of death. Those names do not include rank, birthplace, where they died, or any other information. But since it was impossible to know the order in which the victims of 9/11 died, no such chronology was possible. Clearly, the *New York Times* did not think any other mode of arrangement was appropriate. Contradicting Lutnick's view, the editorial board argued that families could not expect any memorial to "place loved ones in their context, to provide a richer narrative for each individual." Dismissing the sense of family that Lutnick claimed was so much a part of the Cantor Fitzgerald community, and that the rescue worker community held dear, the *New York Times* opined that where people work is not a core part of their identity. "It will always be important to remember the names of the men and women who died at ground zero," the editorial board wrote, "but will it always be important to remember that they worked at Cantor Fitzgerald—to recreate on the memorial the office groupings in which they worked? Death comes to each of us alone, no matter who we work for or who our colleagues are, and it came alone to each of the men and women who died on 9/11. To us, the most respectful acknowledgment we can make to them is to acknowledge them as individuals."[50]

The Struggle for a Buildable Memorial

For a time, it seemed that the LMDC was willing to adopt a simplified version of the compromise position reached by the thirty-two family groups under the leadership of Lutnick, but such an agreement remained out of reach. Eventually, however, the controversy concerning the listing of names assumed less importance for the LMDC, as it continued to struggle to finalize the design of the memorial and reinvigorate a flagging fundraising effort. Neither of these problems would have easy solutions.

The most contentious dispute from a design point of view was Bond's and the LMDC's desire to place the two waterfall voids over the footprints of the two towers. In addition, they wanted to reduce the number of ramps from four to two, one to enter into the memorial space and one to exit, with both being placed in between the footprints. This central area could then contain bathrooms and other infrastructure that Arad didn't account for in his original submission. Initially, Arad and his allies were confused about why Davis Brody Bond would undertake such changes, but it soon became apparent that they were being made to enhance the $160 million memorial museum planned for the site.[51] Although Arad fought the changes, even asking Governor Pataki to intervene, he was ultimately unsuccessful. The LMDC decided to go with the two-ramp design because it was cheaper and seen to be more structurally sound.

By the time the ramps controversy had been sorted out in mid-2006, the cost of building the memorial, which was estimated in January 2004 to be around $350 million, had skyrocketed to close to $1 billion ($672 million for the memorial and $301 million for infrastructure).[52] Some of the additional expense could be attributed to a significant rise in the cost of construction material over the two-plus years since "Reflecting Absence" was introduced to the public. But there was also the additional expense required to build the waterfalls in the void in a way that would allow them to run all year long, even during the frigid New York winter. In addition, there were significant costs associated with shoring up the section of the slurry wall that would be on display in the museum.

Meanwhile, fundraising was stagnant because of the numerous fights taking place over the future of the site. Particularly damaging was the fact that most family groups refused to get behind the fundraising effort until the design was finalized and the names dispute resolved. As a result, the foundation had only managed to raise around $131 million.[53]

The Sciame Report

In May 2006 that Mayor Bloomberg and Governor Pataki formally acknowledged the many problems with the memorial and asked prominent New York builder Frank J. Sciame to get the process under control—specifically to create a memorial plan that could be built in time for the eighth anniversary of the attacks in 2009 within a budget of $500 million. Sciame was asked to preserve the signature elements of "Reflecting Absence" and to adhere to Arad and Walker's "vision" for the memorial. He was given until the end of June (approximately six weeks) to meet with Arad, Walker, Max Bond, and Daniel Libeskind (the group had come to be known collectively as the "Memorial and Master Plan Design Committee"), as well as LMDC officials, the Port Authority, and all major stakeholders (including families, residents, business owners, members of the various memorial committees, and jurors), and come up with a plan to put the project back on track.[54]

In order to accomplish this herculean task, Sciame assembled a pro bono advisory committee of leading architects, builders, business interests, real estate professionals, and engineers. He worked with these advisors, government officials, and the staff of his firm to come up with several alternative plans. Among the most controversial of these were ideas like getting rid of the waterfalls, significantly reducing the landscaping at the plaza level, eliminating the visitor's center, and moving the memorial museum to the Freedom Tower.[55]

On May 31, 2006, Sciame met with the Families Advisory Council to develop a better sense of what they hoped the memorial and museum would accomplish. Early on, it became clear that FAC members were hopeful that Sciame might actually respond to their concerns, but discouraged that they were still rehashing the same issues nearly five years after the attacks. Liz Horwitz, whose son had worked for Cantor Fitzgerald, introduced herself by noting that she had been "coming to FAC meetings for years trying to get something done," while Michael Kuo, whose father had been an engineer with Washington Group International, admitted that "I feel sad all over again just being here."[56] Anthony Gardner agreed: "I can't believe that it's four-and-a-half years ago that we've been attending these meetings and trying to ensure that not only there's an appropriate tribute to our loved ones, but that the story of 9/11 is told and that the authentic remnants of the site and the events are preserved and protected."[57]

After the FAC members introduced themselves, Sciame explained the purpose of his work. He stated that many projects required significant re-examination in order to fit within an acceptable budget. What was different about this project, he noted, was that it "belongs to everyone," and not just a single client. In Sciame's view, the challenge was to make the project feasible. "If we were to align miraculously the budget with the project as designed, it is safe to say that this would be a good memorial, that there would be a museum that would be appropriate where it is now."[58] At this point, FAC member Tom Rogér, the building industry executive whose daughter had been a flight attendant on AA Flight 11, interjected, "I'm not sure you'd have unanimity in this group," to which Allen Horwitz added, "I think you'd be very close to having unanimity on the opposite end of this."[59] For most of the members of the FAC, it was crucial for Sciame to understand that, while families may have been divided over what the memorial ought to be like, there were still many points of agreement. Gardner and Regenhard explained that only a few FAC members endorsed "Reflecting Absence" (Paula Grant Berry, who served on the jury, and Rogér). Lutnick pointed out that the two members who seemed to be the furthest apart in their views (Regenhard and Wolf) could still agree on certain things, especially the need for an "effective contemplative space" that was safe and secure."[60] Lutnick then presented a comprehensive list of things that she believed that most families could agree on:

1. Preservation of, and access to, the box beam columns, the footprints, and the part of the slurry wall required to be saved, based on the historic preservation study;
2. An above-ground memorial so families weren't forced to descend underground to mourn their loved one;
3. A museum to tell the story of 9/11;
4. Nonrandom listing of names;
5. The use of artifacts from the World Trade Center above ground—such as Koenig's *Sphere* and the façade;
6. A tomb that actually contained remains and not a symbolic one;
7. Safety and security;
8. A museum in the memorial quadrant to give context to the memorial.

Yet, instead of encouraging a recognition of shared interests, Lutnick's remarks prompted another family member, Patricia Reilly, to attack Berry. Reilly asserted that the family members who had the most influence in the memorial process did not speak on behalf of the majority of families. Addressing Berry directly, she stated, "You had the honor of being on the jury. You had the honor of being chairperson of the [International Freedom Center]. You are on the Memorial Foundation Board. . . . If you're not passionate one way or the other, then you should listen to the voices of the other family members who don't have that—who don't have the luxury of having all of the connections that you have and have had in this process."[61] Reilly was referring to the fact that Berry was regularly in a position of representing other families—even though she had not been chosen or elected by them to do so—and to her Ivy League and corporate credentials. Berry had an undergraduate degree from Harvard and an MBA from Columbia University, and had held high-level positions in several major media companies. Clearly, Reilly believed that Berry's professional connections, and the ease with which she could interact with other elites in business, politics, and the arts, played a role in her selection for the jury over family members from more modest backgrounds.

Reilly argued that most of the families on the FAC had been forced into a position of being "adversarial" because "everything has been . . . presented to us as a fait accompli. We've always been put a position of being reactionaries. Our input has often been ignored and I think to the detriment of the process. I think you are where you are today, I think we are where we are today because too much of the input of the families, who are just regular people, we're just regular people, you've ignored our input. And if you had listened to it, I think we would be a lot further along."[62]

One thing that everyone in the room could agree on was that Rogér was the only member of the FAC who had never approved of the names compromise that Edie Lutnick brokered, and that this issue was crucial for getting support from the families. Stefan Pryor of the LMDC tried to steer the conversation away from that subject, but family members strongly resisted. They told Sciame directly that resolving the names issue was central to putting the memorial process back on track. The family members reminded him of what they called the original memorial—the missing persons posters, each of which told the story of an individual. Anyone who viewed the posters would see each missing person in human terms. The people on the posters were not just names. They possessed unique identities. Charles Wolf told Sciame that the biggest failing of the Arad design was the names—"it's not just the randomness of the names. It's the indistinction on [sic] the names," he said.[63]

Michael Burke agreed—it was the identities of the victims that told the story of September 11, 2001. He picked a particular example that was not related to the World Trade Center: Bernard Curtis Brown, eleven, who was on AA77, the flight that crashed into the Pentagon. Just seeing that name on a wall does not do justice to the events of 9/11. For Burke, it was absolutely crucial for memorial visitors to understand that Brown was a child from Washington, DC who was embarking on a four-day trip to California with other school-aged children sponsored by the National Geographic Society. To truly understand the depravity of the terrorists, he said, one must realize that "they watched these children and they watched these parents. These were nervous parents. They're kissing their kids good-bye. Call me as soon as you get there. . . . These guys, four guys, know these kids are never getting there. To not express that in a memorial somehow to me is—what is the point of it, to diminish the evil of the day?"[64] Debra Burlingame concurred, noting that a family of four (a mother, a father, and two young children) that died on Flight 77 would be scattered randomly throughout the site, forcing their survivors "to go all over the site looking for their names."[65]

After this exchange, Sciame said that he wasn't likely to be able to resolve the names issue before the deadline set by Bloomberg and Pataki, so he asked to move to the other issues raised by Lutnick. The families then explained that, contrary to the LMDC's assertions, "Reflecting Absence" did not actually preserve the footprints holistically. The area inside the North footprint was broken into several small spaces, including the contemplation room that would house the symbolic vessel and the storage facility where the remains would actually be kept. Several FAC members also expressed concerns that in newer versions of the Arad design the remains would be a part of the museum complex rather than the memorial. They argued that while the bedrock must be preserved, visitors should not be forced to go there, especially to visit the remains. Mary Fetchet suggested that the remains be kept in a private space between the two footprints so that the victims of neither tower are made to seem less important. Patricia Reilly suggested a "mausoleum behind a wall. And behind the wall are the remains and the medical center. . . . It doesn't have to be a tomb. As long as . . . we are looking at it and we're there and we're touching it, our loved ones are behind it. And it's not just . . . empty."[66] Sensing little approval for the symbolic tomb, Rogér told Sciame that it was entirely his idea to create one. Indeed, the original memorial program only specified a repository for unidentified and unclaimed remains, so there was no need to keep the symbolic tomb in the revision.

Gardner, Regenhard, and others also reminded Sciame that the most recent design left bedrock in the north footprint open, making long-term preservation difficult because it would be exposed to the elements. It also created the possibility for a terrorist to throw explosives down into the memorial. This problem was highlighted in a leaked memo from James K. Kallstrom, Governor Pataki's counterterrorism advisor to the LMDC, in which he called for the memorial to be redesigned to "significantly reduce the opportunity for a satchel charge explosive or airborne contaminant dissemination device to be cast, or a suicide attempt to be made into the void."[67]

Halfway through the meeting, Sciame tried to sum up what had been said: First, the families seemed to be in general agreement that the footprints should be preserved in as open a way as possible. Second, the families wanted the remains of their loved ones to be above ground and accessible, possibly in the space between where the two towers once stood. Third, most did not want to have to descend into the museum to reach the remains, both for safety reasons and because paying tribute to the dead would mean entering a museum that would likely charge admission.[68]

The most vocal members of the group agreed with this assessment, but Charles Wolf told Sciame that he was in favor of placing the remains thirty feet below ground because visitors and families would be shielded from the cacophony of the city. In a long critique of Arad's design, Allan Horwitz suggested that Sciame scrap "Reflecting Absence" altogether and chose another design from among the seven other finalists. Sciame politely pointed out that this was not possible.

Those who were concerned about public safety warned that the memorial would be exempt from city building codes. Stefan Pryor from LMDC largely avoided the issue by stating only that the city's building department "has to sign off on the design."[69] Regenhard pointed out that the Port Authority did not have to get a certificate of occupancy, and was only subject to lesser standards outlined in agreements signed between the city and the agency. Again, Sciame seemed to find this conversation unproductive and steered it back to the elements of the memorial design to which the families could meaningfully contribute—particularly whether the memorial should be above ground or thirty feet below ground. Many participants stated once again that preservation of a large section of the slurry wall was never an issue for them—it was Libeskind's idea, not theirs.

As the meeting came to an end, family members expressed hope that Sciame would listen to them and revise the latest version of Arad's plan accordingly. Still, an overriding sense of defeat pervaded. "Do you know how many times I've heard Patricia say this, Tom say this, Mary say this, over and over again to different groups of people, over and over," asked Lee Ielpi. "Do you know how many times we've heard, with all respect, the governor and the mayor say this is it, this is what we're doing? It fails. But this is it and this is what we're doing. It failed again. Ah, we have the solution. It failed. We have another solution. It's failed again. . . . What makes anybody in this room think that anybody's going to believe anything that they say?"[70]

Most family members at the meeting were despondent that they could not have more of an impact on the memorial design process. Debra Burlingame asserted that the families had a special bond with the American people because "our visceral reactions are very similar to theirs" in that all were emotionally affected by the events of 9/11 and needed a memorial that would help them heal.[71]

This view was not universally held by the public, though. If the families thought they had too little say over the design of the 9/11 Memorial, many New Yorkers were growing weary of the constant public presence of victims' relatives who seemed to believe that theirs were the only voices that mattered. The media and the public had become increasingly critical of the families' interventions.[72] At the height of the IFC controversy, for example, one contributor to the *Wired New York* Internet bulletin board offered a cynical take on the role that the families were playing:

> How about two giant sculptures replicating boxes of Kleenex with fountains of water spewing out in every direction from the top to remind us of the tears of all of these eternally grieving family groups. (It must be identified as "Tears of Victim's Families" to everyone to ensure that they all know that no one else really had anything to mourn or to grieve.) . . .
>
> If you have time, you can visit the museum downstairs, but you can only view the exhibits while handcuffed to a family member—this allows the family members to maintain their sense of control, but also lets visitors have someone providing anguished cries of grief or an endless loop of sniffling and tear dabbing through out their experience.
>
> Following the initial first year of opening, the retail center will promote a "victim of the week"—which will allow that victim's family to put up a bigger picture than everyone else and run home movies of that person from exiting the birth canal to the last birthday party. Before you leave the Kiosk area, you will have to pass through the taxation center, where you will be forced to pay a 10% NYPD/FDNY tax on any purchases—because their lives were more valuable and, if the civilian victim's families are going to get any money for their loved ones being at work on 9/11, then the FDNY and NYPD certainly deserve more money for being at work—because they were working with an expired contract.
>
> Thank you for visiting the "New York Wailing Wall." And remember, the Victims' Families love New York more than you and are more American than apple pie.[73]

Resolution

On June 15, 2006, Sciame met with Mayor Bloomberg and Governor Pataki to discuss the progress of his review. On June 20, his report, which detailed significant design changes, was released to the public for comment. It was clear that Sciame had listened to, and taken seriously, the views of the families. Most importantly, Sciame moved the names of the victims up to street level, placing them on parapets that surrounded the waterfalls in the voids (which were retained, along with Walker's landscaping). This change would eliminate a major source of anger among dissenting families and improve accessibility. As a result of this change, the ramps that had caused so much strife were eliminated and some of the below-ground galleries of the memorial were removed, dramatically increasing the space available for the museum. The underground views of the waterfalls and the contemplative spaces were preserved—at least for the moment. These aspects of the Sciame plan would be changed again in subsequent revisions for cost and aesthetic reasons.

The second most important change was that the entrances of the below-grade Memorial Hall and Memorial Museum were consolidated into a single entry point within the Snohetta-designed Visitor Orientation and Entrance Center (VOEC). This redesign would not only make the visitor experience more seamless, it would significantly reduce costs by eliminating redundant facilities—especially in terms of security. As described in Sciame's report, "The VOEC will serve as an iconic presence for the Memorial Museum upon arrival at the site. Visitors will enter the VOEC and choose a) to ascend to a site-wide orientation experience and education exhibits or b) to descend to the Memorial Hall (located between the two pools) and the Memorial Museum. Here within the Memorial Hall, visitors will have the opportunity to view the waterfalls of both the north and south pools in a place for gathering and reflecting below grade."[74] The museum would include the slurry wall and box-beam column remnants.

These changes had a profound impact on the meaning of the memorial at Ground Zero. Although numerous alterations were made to the revised plan before the memorial and museum were built (including sealing the waterfalls so that they were no longer visible from the interior), the general scheme would remain the same. Crucially, the large contemplative spaces that once dominated Arad's memorial were replaced with an expositive museum that left little room for interpretation. The story of 9/11 would be interpreted by museum planners, not visitors.[75] Further, the museum and memorial would no longer be separate entities—they became intertwined both physically and administratively and would be thought of as a single unit. To reflect this consolidation, the World Trade Center Memorial Foundation announced in August 2007 that it was changing its name to the National September 11 Memorial & Museum at the World Trade Center. It also underscored the foundation leadership's desire to focus on all three September 11 sites—Shanksville, the Pentagon, and the World Trade Center—not just lower Manhattan.[76]

Surprisingly, Sciame did not address one of the most sensitive aspects of the memorial design: where to house the unidentified and unclaimed human remains. Neither the symbolic tomb nor the OCME facility appeared in the report and no suggestions were made on replacing them. Sciame left it to the foundation, LMDC, and design team to decide; he never explained why he declined to weigh in on this issue.

According to the post-Sciame design revision released by the LMDC on September 29, 2006, unclaimed and unidentified remains were to be stored in a separate facility behind a large wall at bedrock between the footprints of the two towers. Visitors would be able stand before the wall, but could not access or see the remains repository itself. Relatives would be able to access the repository either through a door at bedrock or through an elevator from the surface directly down to bedrock and into a special family room adjacent to the repository behind the wall.[77] The legitimacy of this decision, and the extent to which it was made with the input of families, would become yet another sources of controversy.

Resolution of the Names Controversy

Although Sciame brought the names of the victims above ground, he did not make any recommendations about how they ought to be listed. When Mayor Michael Bloomberg became chairman of the Memorial Foundation in October 2006, however, he promptly reaffirmed his desire for the names to be listed randomly as Arad intended.[78] This angered the majority of families, who then refused to support fundraising efforts for the memorial, which continued to hurt the ability of the foundation to raise money.

In December 2006, however, Bloomberg rather abruptly changed direction. He made a deal with Stephen Cassidy, head of the Uniformed

Firefighters Union, who wanted all firefighters to be listed together by unit, with rank noted. Under this deal, the names of rescue workers (firefighters, police offers, and EMS) would be identified by affiliation and grouped by company or precinct around the South Tower pool. Bloomberg would not budge on the rank issue, though, and Cassidy was forced to accept names only. Lutnick reports that many Pentagon families found this concession deeply troubling since rank was central to the identity of military personnel. Also listed in this area would be the victims who died on board UA175 when it hit the South Tower, those who died on the planes that crashed into the Pentagon and the field in Shanksville, the victims of the 1993 World Trade Center bombings, the victims who worked in the South Tower, and those victims whose exact location could not be determined.[79]

Around the North Tower parapet, the names of the victims who died in that tower would be listed, along with the passengers and crew who died aboard AA11 when it crashed into that building. In a move that deeply angered Lutnick and many other civilian families, while employees could be grouped together under Arad's "meaningful adjacencies" plan (in which each family could chose two people to be listed next to their loved one), companies, organizations, and floor numbers would not be listed. Thus, civilian names—unlike the firefighters and those who died on the airplanes—would appear to be random even if a large group of coworkers or people who shared other relationships were listed together.[80] According to Lutnick, the lack of contextual information such as age and affiliation would make it difficult for visitors to establish an emotional connection with civilian victims. This plan not only failed to satisfy many civilian families, it also created problems for the designers, requiring a specially designed computational algorithm, handmade mockups, and sometimes arbitrary human decisions. In the end, Lutnick was able to group the 715 Cantor Fitzgerald employees and contractors plus the 61 Euro Brokers employees, but only under the dictates of the meaningful adjacencies plan and a $10 million donation to the Memorial Foundation by her brother Howard.[81] For most other civilians, the final decisions about name placement were made by the LMDC and Arad.

Within a few months of taking control of the Memorial Foundation, Bloomberg had raised over $100 million and hoped to have the memorial area of the complex complete by the time he was supposed to leave office in 2009.[82] Activist families and their supporters balked at his control over the foundation. For starters, they were concerned about potential conflicts of interest. Bloomberg was asking for donations from wealthy individuals and huge corporations while he was mayor of the financial capital of the world. In addition to the money from Cantor Fitzgerald, the foundation received significant donations from Deutsche Bank ($15 million); American Express, Goldman Sachs, Merrill Lynch, Con Ed, and Verizon ($10 million each); Larry Silverstein ($5 million); the owners of several New York area sports teams (in the range of $500,000 to $1 million); plus many others.[83] "The conflicts of interest as well as the obvious political reasons behind the appointments just fly in the face of the 'open and democratic process' that they have always claimed to [be] behind the memorial process. I must have missed the memo that changed the process from a democracy to a monarchy," wrote Dennis McKeon in an editorial published by *Downtown Express*.[84] Curiously, it seems that Bloomberg predicted his ascension to the head of the foundation. In a speech to business leaders just a few months after September 11 attacks, Bloomberg stated that, although he could not envision what the memorial to the victims would look like, he could most certainly envision the process: "everybody yelling and screaming for a number of years and then somebody taking charge and just doing it."[85]

More fundamentally, activist families were angry that that Bloomberg did not seem to empathize with them (see chapter 2). In Bloomberg's view, the goal of memorialization should be to build a better future and not to dwell on the past. Activist families, on the other hand, wanted a memorial and museum that gave precedence to the individual stories of their loved ones and reflected what they described as "history not 'his story'."[86] A memorial, in their view, should not primarily serve economic interests, drive tourism, enhance a residential neighborhood, or burnish the reputations of politicians and influential actors. A well-designed and well-built memorial would do all of these things, but they were of secondary importance. As Anthony Gardner told me, "the best things about the memorial and museum stem from the families' engagement. That's what is going to resonate with the visitors and that's what they are going to remember. We always had the long term in mind. We weren't just being angry or obstructionist and we weren't angry or irrational. We weren't just in it for the sake of our own loss, but for the sake of history. I don't apologize for what we had to do to achieve a meaningful memorial and museum."[87]

Negotiations

Despite all of his political power and financial clout, Bloomberg could not build the memorial and museum on his own; he had to work with the architect, the Port Authority, which was responsible for building the complex, and Bovis Lend Lease, the company the Port Authority had subcontracted to do the actual work. Negotiations among these parties, plus the myriad other interest groups that had, or wanted to have, a say in what the memorial and museum would be, were difficult. The complexity of the rebuilding of the World Trade Center site, in which each step in the process was dependent on several others, made the talks even more unruly. The memorial and museum, to take but one example, was to be built directly on top of the new transportation hub's PATH train mezzanine. As such, it could not be built until the roof of the PATH train infrastructure was in place, and the roof design of the PATH train could not be finalized until engineers knew exactly what kind of weight load and distribution they would have to plan for. But, as stated in a Port Authority report on the challenges of rebuilding the site

> the Memorial Foundation and Port Authority had been negotiating for months . . . and could not reach agreement on the number of trees that could be accommodated on the northeast corner of the Memorial quadrant. While this may sound like a relatively trivial design element in the scheme of things, given that each tree—with the tree itself, soil and planter—weights 150 tons, that means that approximately 7,500 tons (50–60 trees to be placed on the NE corner of the Plaza) had to be factored into the design of the Hub's roof support—approximately the equivalent weight of 180 fully loaded tractor trailers or half the weight of the Brooklyn Bridge.[88]

If the decision about how many trees ought to be included in the memorial was this complex, one can only imagine how challenging it was to make choices about more pressing issues such as transportation needs, security, safety, critical infrastructure, design, and cost containment. All told, there were "19 public agencies, two private developers, 101 different contractors and subcontractors and 33 different designers, architects and consulting firms all in charge of one element of the project or another."[89] Chaos seemed to be the only possible outcome.

Indeed, this was essentially the Port Authority's conclusion when it finally admitted publically in June 2008 that the effort to rebuild the World Trade Center lacked a "functional decision-making model."[90] In a rare moment of candor, the Port Authority noted that it "could not answer with confidence" four of the most basic questions that any project manager ought to be able address: "what we are building, who was building it, when it would be built, and for how much."[91] In its June assessment, the Port Authority identified fifteen issues that had to be resolved with honesty and "clear-eyed" analysis "before we could answer these questions with certainty."[92] It also insisted that all stakeholders had to work together to "get the rebuilding program to a level of certainty and control so that schedules and budgets reflect the construction reality on the ground instead of politically- or emotionally-driven promises."[93]

On October 2, 2008, the Port Authority released what it called "A Roadmap Forward," in which it addressed each of the fifteen issues identified in the June assessment, promised to be more responsible, honest, and accountable to the public in the future, and issued revised deadlines and budgets for future construction. While the Port Authority expected skepticism about the findings of the report, it claimed that the new plan was different from past ones in that it was realistic and incorporated "construction realities that had not been fully understood until now," risk calculations, and milestones that would force the Port Authority to be open to public scrutiny.[94]

Perhaps the most important result of the assessment was the decision of the Port Authority to prioritize the opening of the 9/11 Memorial in 2011, in time for the tenth anniversary of the attacks. The dependence of the memorial and museum on the construction of transportation infrastructure, as well as their reliance on a dysfunctional chain of command for the rebuilding effort, meant that it was difficult to foresee a completion date for the project. In order to alleviate the first problem, the Port Authority decided to build the roof of the transportation hub's PATH Mezzanine before actually building the PATH platform and other critical infrastructure. Once the roof was complete, then construction of the memorial could begin. In order to alleviate the second problem, the Port Authority took complete control of the construction process, leaving it to the Memorial Foundation to deliver construction plans to the Port Authority and their contractors. The Port Authority also made important design and construction choices regarding the roof of Santiago Calatrava's transportation hub entrance and the #1 subway line, and resolved a seven-year-old land claim dispute that prevented the construction of the Vehicle Security Center, a crucial aspect of the site's overall security plan.[95]

With the Sciame plan accepted and the Port Authority taking ownership of the delays and ineptitude that had marred the reconstruction effort up to this point, it seemed as if the construction of the memorial and museum, as well as One World Trade Center and other buildings, was finally on track. Sadly for everyone involved, however, controversy surrounding human remains would not go away.

7

New Finds

City workers, the FDNY, and the private contractors charged with cleaning up the World Trade Center site understood very well that the ceremony held on May 30, 2002 to mark the "official" end of the recovery effort was largely symbolic. A tremendous amount of work still needed to be done outside of the Pit to clean up debris and search for human remains, particularly in the buildings around the site that were heavily damaged, but remained standing, after the attacks: 90 West Street, 130 Cedar Street, and the Deutsche Bank/Banker's Trust Building at 130 Liberty Street. Although these buildings had been searched in a cursory way shortly after the September 11 attacks, they had not been thoroughly scoured for human remains. Indeed, searches of the roofs and upper floors of the West and Cedar Street buildings throughout the summer of 2002 turned up numerous bone fragments that had been missed the first time around.[1]

Some families, especially those already mobilized in protest of the long-term storage of World Trade Center debris at Fresh Kills, or on issues related to the memorialization and reconstruction efforts, began to question the quality and efficacy of these searches when construction workers rehabilitating the 90 West Street building for residential use discovered additional remains on September 8, 2003. Nine pieces of bone and tissue were found on exterior scaffolding that had been put into place when the renovation work originally started in 2000.[2] While there was no large-scale effort to publicize what some families considered to be the shortcomings of the recovery effort, the stage was set for a larger battle over the discovery of human remains at the Deutsche Bank Building, when it was finally being readied for demolition in September 2005, four years after the terrorist attacks on the World Trade Center.

The Deutsche Bank Building

In August 2005, the LMDC purchased 130 Liberty Street from Deutsche Bank in order to demolish it and ensure that the redevelopment of the World Trade Center site would not be delayed.[3] The forty-one-story building, which sat directly across the street from the South Tower and was heavily damaged by debris from the towers as they collapsed, had remained shrouded in black netting for years as Deutsche Bank and its insurers fought a very public legal battle over whether the building ought to be repaired or torn down. Once the acquisition was complete, the LMDC had to formulate a deconstruction plan to be approved by the EPA and other regulators, and hire private contractors to clear the structure of toxic materials such as asbestos, dioxin, lead, silica, quartz, chromium, and manganese, among other substances, so that it could be safely deconstructed.[4]

While the FDNY had conducted a search for human remains during the early days of the recovery effort in 2001, and then again in June 2002, they were more concerned about recovering large, visible pieces of bone and tissue than smaller, pulverized fragments or those hidden away in difficult-to-reach areas. According to an FDNY spokesman, firefighters did a visual inspection only—"we didn't dismantle, we didn't sift through. We didn't look in air ducts, that kind of thing. Nothing was discovered. We never went back in there."[5] The same was true for other areas adjacent to the World Trade Center site. As environmental abatement workers were sifting through the ballast (gravel) that held the roof membrane on top of the Deutsche Bank Building in mid-September 2005, they found a few small objects, in the range of one to two inches, that appeared to be human remains.[6] These objects were turned over to the medical examiner's office, where they were confirmed to be human in origin and potentially identifiable.[7]

It is important to keep in mind that these discoveries were taking place amid raging debates about the future of the memorial, the controversy over the International Freedom Center, and the politics of the redevelopment of the site. As such, the travails of the Deutsche Bank Building and the discovery of human remains on its roof fed into, and were amplified by, the ongoing dysfunction at the World Trade Center site. The families that were directly engaged in activism were primed to interpret any unsettling news through the prism of their ongoing struggles to influence the development of the site and the memorial to their loved ones. Relatives active in World Trade Center Families for Proper Burial (which was advocating for the removal of World Trade Center debris that might contain minute human remains from Fresh Kills) were also likely to view the remains recovered from the Deutsche Bank Building as potential traces of their sons, daughters, wives, husbands, brothers, or sisters. Their reaction to the discovery of new remains four-and-a-half years after the collapse of the twin towers can only be understood in this context.

After the discovery of remains at the Deutsche Bank Building, activist families demanded that demolition be halted so that a more thorough search of the building could be conducted.[8] The Lower Manhattan Development Corporation and the city assured families and the public that it was taking every necessary step to ensure that all human remains would be recovered from the site, but stopped short of promising to halt demolition to accommodate an independent search effort.[9]

The situation took a significant turn for the worse in early April 2006 when the *New York Times* and other newspapers reported that demolition workers had found an additional seventy-four bone fragments in the ballast that had been supposedly thoroughly searched by environmental abatement technicians in September 2005. In response to these new finds, two firefighters spent nearly two weeks on the roof meticulously raking through the ballast, spreading it out flat and searching through it on hands and knees. They found several additional remains.[10] While the LMDC attributed these additional finds to the "rigorous protocols" handed down to the LMDC by the OCME, and the great care taken by civilian construction workers to thoroughly search for remains before the dismantling of the building, skeptical family members were beginning to question the competence of the LMDC and its contractors. They were angry, and often incredulous, that remains were still undiscovered four-and-a-half years after the attacks.[11]

On April 8, 2006, representatives of several family activist groups released a scathing statement to the media, chastising the LMDC for using "construction workers" to conduct searches for human remains in the Deutsche Bank Building and withholding information about the bone fragments for several days after their discovery. They called on Mayor Bloomberg to intervene and get a team of experts from the OCME to oversee the recovery process. Convinced that other searches for human remains near the World Trade Center site were fundamentally flawed, these family activists also demanded that previously searched sites be investigated again by OCME scientists to ensure that no remains had been missed.[12] Calling the validity of the entire recovery operation into question, the activists wrote:

> In terms of expertise, caring and honesty, the Office of the New York City Medical Examiner has proved its integrity. The families, who suffered such horrid losses on September 11, 2001, have come to trust this department and to honor their dedication to careful, thorough and experienced investigation. While we have no objection to the presence of FDNY and NYPD personnel, and even welcome these additional 'eyes,' we ask that the trained personnel from the Medical Examiner's office be given jurisdiction over this crime scene investigation. . . . The LMDC must be reminded that this is a crime scene. . . . While Firefighters and Police may serve in an honorary function, the sacred & scientific retrieval of the human remains of WTC victims must be squarely in the hands of seasoned professionals of the NYC Medical Examiner's Office. Economic considerations and expediency cannot substitute for the dignity and professionalism that the victims of 9/11 and their families deserve.[13]

While OCME officials must have appreciated the initial vote of confidence from families, they did not agree that the recovery effort needed to be taken over by their office, or that the discovery of additional small remains was problematic. OCME spokeswoman Ellen Borakove stated that her agency approved of the work being done by the LMDC contractors and praised construction workers for their attention to detail in recovering even the smallest bits of human remains from the site, including an additional 300 "tiny" bone fragments, some as small as a sixteenth of an inch—from the roof ballast in mid-April.[14] This lack of receptivity to the concerns of the activist family members seems to have been the first step in their disillusionment with the OCME personnel who had replaced the scientific team that worked on the World Trade Center case from 2001 to 2004.[15]

The same day, the Skyscraper Safety Campaign, spearheaded by Sally Regenhard and Monica Gabrielle, issued a follow-up to the previous week's family press release, cosigned by Jim McCaffrey, Rosaleen Tallon, and

Dennis McKeon, expressing shock and horror over the recovery of additional remains from the roof of the Deutsche Bank Building and the failure of the LMDC to inform families immediately about this discovery. The activists chastised the LMDC for ignoring their requests for additional information and for detailed protocols that explained exactly what was being done to ensure the scientific recovery of remains.[16] The LMDC did release a September 26, 2005 memo detailing what to do in the event that additional remains or personal effects were discovered, but the instructions essentially boiled down to: carefully recover them, take note of the circumstances in which they were discovered, and call us.[17]

For these family members, the actions of the LMDC were so egregious, that they called for the agency to "relinquish control of Ground Zero." They went on to demand federal intervention and asked President Bush to assign the Joint POW/MIA Accounting Command (JPAC), the wing of the U.S. military tasked with recovering and identifying missing soldiers, to assist in the recovery effort at the Deutsche Bank Building and in other areas around the World Trade Center site that they felt had been inadequately searched. Specifically, they demanded that a grid system be established and that every inch of debris be screened for potential human remains. They concluded by noting that the LMDC had not followed protocols like those used by JPAC and that it was time for "professionalism" at the World Trade Center site. "It has been said that our First Responders and loved ones were the first soldiers in the war on terror. Now it is time for the federal government to come to the aid of our valiant lost loved ones."[18]

In the weeks after their demand, New York senators Charles Schumer and Hillary Clinton made similar requests to the Pentagon on behalf of the families. No action was taken, however, because federal law required local officials to request help before it could be provided. The families sought such a request from Mayor Bloomberg and Governor Pataki, but neither was willing to make the call to Washington. They also requested support from Attorney General Eliot Spitzer.[19] The mayor's office refused to budge. Patrick J. Brennan, who was commissioner of the Mayor's Community Assistance Unit, praised the LMDC, the OCME, the FDNY, and volunteers from previous searches for their hard work and adherence to existing protocols at the Deutsche Bank site. He deflected calls for JPAC to be brought in and instead assured Regenhard and those who had advocated for outside intervention that everything was under control at the site.[20]

LMDC officials stated that they could not be blamed for any missteps made during the initial recovery effort. They argued that they inherited a problem when they purchased the Deutsche Bank Building and that they were doing the best possible job of recovering remains under the circumstances.[21] LMDC family outreach coordinator Anthoula Katsimatides sought to address the concerns raised in the initial Skyscraper Safety Campaign press release. In a letter to families, she assured relatives that protocols for the recovery of remains at the Deutsche Bank Building were "implemented in conjunction with" the OCME and that the effort was being carried out under the supervision of a professional anthropologist, Larry McKee, from the engineering and construction management company TRC Solutions. She also noted that a medical legal investigator would be posted at the site to "ensure that remains will be identified and handled in a way that is respectful and sensitive to the needs of the families."[22]

Throughout late April and May 2006, the local media took note of this story as part of its ongoing effort to examine the human remains issue at the World Trade Center site.[23] One story described the effort to tear down the building as "hexed," and the problems there as continuing to "fester and grow."[24] These references alluded to more than the discovery of human remains at the site so long after the attacks.[25] They also invoked U.S. Occupational Health and Safety Administration (OSHA) concerns about worker safety at the site and the U.S. Environmental Protection Agency's decision to shut down the site in order to resolve problems with the environmental impact of the demolition effort. Such challenges not only slowed the recovery effort and demolition work, they also undermined confidence in the LMDC among many families and residents of lower Manhattan.[26]

As the activist families' desperation grew about the fate of remains at the Deutsche Bank Building, they placed a call to Richard Gould, the forensic archaeologist from Brown University who suggested in 2001 that human remains would likely be found well outside the World Trade Center site. Although he did not find human remains at the time, the subsequent discovery of bone fragments on the rooftops of buildings adjacent to the site confirmed his hypothesis. Gould and his colleague Anne Marie Mires immediately agreed to help and sent a letter to Mayor Bloomberg telling him that "information we are currently receiving from different authorities and sources indicates that the recovery work at the Deutsche Bank Building does not meet minimum standards of forensic or archaeological science for this kind of important work."[27] They argued that to be legally valid, the recovery of human remains and other effects must be done by "credentialed specialists . . . who not only perform scientifically-controlled work at the scene but are familiar with chain-of-custody and other medico/legal procedures along with health and safety issues." Gould and Mires further stated that construction workers were ill equipped for such work and that their understanding was that unscientific methods were being used, without appropriate mapping, sieving, controlled photography, and detailed recording of descriptions of remains. Once again, Gould offered his organization's services to the city at no charge and stated that, even if no remains were recovered, scientifically valid efforts to locate them could prove beneficial to distraught family members.[28] Gould also participated in protests and rallies organized by families seeking greater accessibility to the Deutsche Bank site.

OCME officials and others involved in the recovery effort were dismayed by these efforts to intervene in the recovery effort and the use of media. They believed that their procedures and protocols were as rigorous as JPAC's. Indeed, Bradley Adams, head of the OCME Forensic Anthropology Unit, noted that he had worked at JPAC before coming to the OCME.[29] Larry McKee even wrote to Gould to let him know that his concerns were based on rumor and hearsay and not on the actual workings of the recovery effort.[30] In a follow-up e-mail, Gould withdrew some of his original claims, and instead opined that OCME's unwillingness to discuss the details of the recovery with concerned family members undermined its efforts. He wrote:

> These people [the family members who had contacted him] desperately need reassurance that the recovery effort is being performed up to fully professional standards, and they and I both cannot understand why OCME is being so secretive about this. We all know that while it is not appropriate to discuss the evidence or the findings of your work publicly, it is ok—and I would even say, essential—to be open about the conduct of your work. . . . I am afraid it took some harsh words to persuade OCME to permit an unbiased site visit by an outside observer to make such an assessment possible.[31]

In order to assuage families, and to rebut the claims made by critics, the OCME allowed Gould to come to the Deutsche Bank Building for a day to observe the recovery effort. The LMDC insisted that Gould submit to a number of requirements before accepting the invitation, including various medical tests and completion of a two-hour asbestos awareness training session. In addition, the LMDC also demanded that Gould be covered by worker's compensation, "provide proof of adequate general liability and environmental insurance naming LMDC as a named insured in a matter satisfactory to LMDC," and sign a release promising not to file legal claims against LMDC or its affiliates as a result of the visit.[32]

After completing all of the requirements, Gould visited the Deutsche Bank Building on August 8, 2006. Upon arrival, it immediately became clear to Gould that the OCME, the LMDC, and their affiliates were doing a good job of recovering remains from the building. Since the initial finds, they had implemented a scientifically sound plan, which included sieving, carefully cleaning and preparing each remain recovered, recording important information about each fragment located, and working toward the goal of 100 percent recovery. He concluded that "the conduct of the operation I saw that day was impeccable and conformed to the best standards of good archaeological science" and is "the best possible under the circumstances."[33] In his view, the protocols in place would ensure a 90 to 95 percent recovery of human remains at the site, which is an excellent yield for a disaster site. He further noted that appropriate safety mechanisms were in place and that the operation was adequately staffed. In his report, Gould was careful to note that he was evaluating the recovery effort as it was taking place in August 2006, and not before, when protocols and methods did not appear to be as well developed. Indeed, in a letter to Ellen Borakove accompanying his final report, he explained that he chose not to dwell on the question of why it took so long for scientifically valid methods to be put into place after the initial discoveries of human remains at the Deutsche Bank Building in 2002 "because the damage that this caused cannot be undone."[34]

Gould concluded his report with some advice for the OCME regarding its relations with concerned 9/11 families: tell them what is going on, at least at an operational level, lest they "speculate and jump to conclusions."[35] "I think some of the misunderstandings during the Deutsche Bank Building recovery operations could have been avoided by timely bulletins from the OCME about the excellent work taking place there, once it began. . . . The 9/11 families, in particular, would have been reassured to know more about the fine work being done on their behalf, and I am sorry that they are only finding out about this now."[36]

Gould's report was not enough to reassure Regenhard and other family members that the LMDC and OCME would continue to use such rigorous methods to search for remains at other locations in the vicinity of the World Trade Center that they felt had never been properly searched.[37] Families were also upset that the LMDC declared the Deutsche Bank Building free of human remains after a search of the roof and mechanical and ventilation equipment that vented to the roof, but not other areas of the building, most notably around the "gash" in the building's exterior caused by projectiles from the collapsing South Tower.[38] In their view, the unwillingness to search these areas was emblematic of a lack of real concern for the families of the missing. In a letter from Al and Sally Regenhard to Ruth Simmons, the president of Brown University, thanking her for supporting Gould's work, they wrote:

> At last count, there were still 1,152 people who disappeared in the World Trade Center and are still unaccounted for. . . . Even for [those families] who have received "remains" many of these are only partial and microscopic, which essentially means that human remains from a vast number of 9/11 victims may still remain at GZ [Ground Zero] & elsewhere. We are hoping to see that terrible number shrink in the near future as the improved recovery work at the Deutsche Bank Building and in other nearby locations in the perimeter of GZ proceeds. We also know, however, that many remains that could have been identified earlier were lost due to less-than-adequate recovery methods, and that is a heartbreaking reality and failure of the system that we now have to live with every day of our lives. Anyone who has experienced this kind of loss suffers a double blow. First, there is the virtual certainty of the loss, based on circumstantial information. That is bad enough, but the second blow, the never-ending ache caused by the uncertainty of just where those human remains actually are. It's hard to begin to grieve adequately without those remains.[39]

All in all, it looked as if the human remains issue had reached settlement in lower Manhattan. The LMDC and OCME were vindicated by Gould's report, but the families continued to believe that their complaints about the fundamental problems with the remains recovery effort were falling on deaf ears. These families were convinced that more remains were waiting to be discovered near the World Trade Center site, but nobody seemed to be listening to them. Over 700 human remains from those who died on September 11 had been recovered from the Deutsche Bank Building. Few of those remains, though, were larger than a penny.[40]

Below Ground

As the summer of 2006 turned into autumn, construction on the infrastructure for the rebuilt World Trade Center site was beginning. One of the first tasks facing workers was the need to clear out the manholes that had become clogged with debris during the collapse of the twin towers and the subsequent recovery effort. These areas, many of which were within the bounds of the sixteen-acre site, had only been searched in a cursory manner, and many had been paved over to allow for trucks and other heavy equipment to get through to the Pit in the first few months after the attacks. Among the places being cleaned were two manholes underneath a paved road that had been covered up since 2002. On October 20, 2006, Con Ed workers used powerful vacuums to suck out debris. They loaded as much as they could into a truck and drove it back to a Con Ed operations center a few blocks away.

Hours later, a contractor working with Con Ed was loading additional material into another truck when he noticed two eight- to ten-inch objects sticking out of the rubble. Curious about what they were, he took a closer look. Confirming many 9/11 families' worst nightmares, they turned out to be human bones. The worker immediately notified his superiors, setting off a flurry of activity to prevent the debris that had already been trucked away from being dumped in a New Jersey landfill.[41] Subsequent investigation of the area and the debris that had already been trucked off to Con Ed's facility revealed more than 100 additional bones, as well as numerous personal items, including two wallets. Unlike the miniscule fragments recovered from the ballast of the Deutsche Bank Building, many of these remains were large and recognizable as specific human bones to trained anthropologists. There could be no avoiding the fact that sections of the site were not adequately searched during the initial recovery effort.

Families were dismayed by this news. Even relatives who were not regularly involved in disputes about human remains found the discoveries unsettling. Monica Iken, for instance, told reporters that "the reality is that there are remains down there. You never want to think that your loved one is in a Dumpster somewhere. It's hard to hear that the [recovery] job wasn't done."[42]

For families engaged in human remains activism, these new finds were confirmation that local authorities were incapable of leading the search and should be replaced by specialists from JPAC. They were incensed that the discovery was made by chance by untrained construction workers and that nobody in charge anticipated that human remains and other World Trade Center material would be found in manholes that were covered up in the wake of the attacks. In a letter to families on October 19, 2006, Dennis McKeon from *Where to Turn* linked the failure to properly conduct the search for human remains to a more general concern about the corruption of the people leading the redevelopment effort. "It is a crime that the powers that be are more concerned about fundraising [for the memorial]," he wrote, "than they are about a proper search for remains."[43]

City officials sought to downplay the significance of the finds while, at the same time, promising to do more to ensure that any additional remains would be recovered. At an emergency meeting on Friday, October 20, Mayor Bloomberg vowed to determine why these remains were not discovered during the initial recovery effort in 2001 and 2002 and to lay out a comprehensive plan to discover whether any other remains could be found.[44] At the same time, Bloomberg told listeners of his weekly radio program that the workers who participated in the initial recovery effort "did the best they could. You can't be perfect, unfortunately, when it's something this big." He later admitted, "whether or not two years from now or during construction somebody finds something else, you just don't know."[45]

While Bloomberg was busy explaining that the discovery of additional remains was a result of the enormity of the site, his deputy Edward Skyler hunkered down with city officials, Port Authority personnel, LMDC representatives, and utility company executives to develop a plan to search all remaining manholes and other similar spaces for World Trade Center debris and human remains. They decided that Con Ed and Verizon Telecommunications would immediately search any manholes that were identified in a comprehensive review of the area. OCME personnel would be on hand to examine the contents of these spaces using the same rigorous protocols that it eventually put into place at the Deutsche Bank Building.[46]

By Sunday, additional finds were being made in newly searched manholes in the vicinity of West Street, including more than a dozen human bones (many of which were several inches long) and numerous personal effects.[47] While newspapers reported that uniformed personnel and civilian construction workers had revealed toward the end of the official recovery effort in May 2002 that the ground underneath the access road adjacent to West Street had not been properly searched, the city did not formally comment on these accusations. Officials continued to claim that one could not expect a sixteen-acre site to have been meticulously searched for even the tiniest fragment of human remains. According to a report in the *New York Daily News*, the Department of Design and Construction considered the option of doing a search of manholes and pockets created by flying debris spearing into the ground, but determined that the two weeks necessary to do so was too long and declared the area free of human remains.[48] Further, the search would have taken years if all rubble from the World Trade Center was run through one-sixteenth-inch mesh. The people in charge of the recovery effort had a legitimate interest in clearing the site quickly, especially in the first few months of the effort, and therefore did not formulate a plan that included rigorous screening.

Activist families were quick to pounce on this new cache of remains as further evidence that the city was incapable of conducting a proper scientific search for their loved ones. Rather than focus just on the OCME's current protocol, they took the long view, and considered the discoveries as part of an extended history of mismanagement and incompetence on the part of city officials. As such, they demanded once again that the city relinquish control of the recovery effort and call in experts from JPAC. They were joined once more by senators Schumer and Clinton, who wrote a letter to Mayor Bloomberg asking him to call in the military's recovery team. "Despite the considerable, sincere, and often excellent efforts of those involved to locate and identify all human remains in lower Manhattan . . . more resources and expertise are needed to complete this essential task. We urge you to echo (the families') calls for . . . JPAC or other appropriate agencies to assist in the investigation."[49]

The mayor's office rebuffed these demands, stating that while there had certainly been pressure to clean up lower Manhattan quickly, this pressure did not prevent recovery workers from "doing everything that was appropriate at the time."[50] The mayor's family outreach commissioner, Patrick Brennan, further promised families that the city had the best possible people working to recover additional remains and that "we have two former members of that command (JPAC) on the staff of the Medical Examiner's

Office who have advised us that the command's expertise is in searching for missing soldiers, not bone fragments."[51]

Families were outraged by claims that the best people were on the job—they argued that this was no task for utility workers and individuals covering up for past mistakes, but rather the trained anthropologists and archaeologists who had not been involved in the previous efforts. This time, they went even further, calling on Attorney General Spitzer to investigate the recovery effort. They also demanded that construction be halted until a "total coordinated search of the entire area" could be carried out.[52] Despite these demands, city officials, including Mayor Bloomberg, reaffirmed that construction, which had been delayed for so long by bureaucratic infighting and money woes, would continue unabated.[53] Sally Regenhard accused the city of "focusing on the rebuilding rather than the sanctity of human life and death," and families stated that they were mulling over the option of asking for a federal investigation into the treatment of human remains at the World Trade Center site and Fresh Kills since September 11.[54] For these families, the failure to fully investigate the bungling of the recovery effort—which they likened to obstructing an ongoing murder investigation—was no less a scandal than an improper investigation of the 9/11 attacks themselves would have been.[55]

Families also criticized what they perceived as the city's overly narrow plan to concentrate primarily on seven manholes and three service boxes (more shallow spaces than manholes which were used to access utility lines), as well as Verizon boxes, transit boxes, and sewers around the West Street service road, rather than thinking about lower Manhattan more holistically. A release from Put It Above Ground directly challenged the city's technical expertise:

> They are still going at this the wrong way. They are developing a plan from this site and working their way out. They [the city] need to determine the entire search area (Ground Zero and surrounding buildings) and work their way in from the outer perimeter. And they need to realize that the hundreds of thousands of tons of pressure that were brought down to street level when the buildings collapsed could have pushed remains through every available outlet for miles. All tunnels, sewers, conduits, and other possible exit strategies [sic; probably subways] need to be searched. . . . JPAC could create such a plan. But as we all know they are refusing the help. We need to keep the pressure on.[56]

In many of their public messages, they also called for a new search at Fresh Kills, since they now believed that the city was not capable of conducting a competent search for human remains.[57] As the recovery effort continued throughout the week, additional remains were uncovered, including reports of hair and scalp as well as many other large bones. The activist families' frustration mounted. They hoped to be invited to meetings taking place among city officials and representatives of FDNY, NYPD, DEP, Verizon, Con Ed, and the LMDC, but they were not.[58]

The Burney-Maikish Report

By October 27, the total number of remains that had been recovered from the past week stood at around 200. That day, David J. Burney, commissioner of the Department of Design and Construction, and Charles Maikish, executive director of the Lower Manhattan Construction Coordination Center (LMCCC), an offshoot of the LMDC, also formally submitted their "Report and Recommendations on Additional Searches for Human Remains at and in the Vicinity of the World Trade Center" to Deputy Mayor Edward Skyler. This report outlined the events that led to the resumption of the search for human remains at the World Trade Center site and presented a plan. In order to create this plan, Burney and Maikish convened representatives of all agencies and organizations involved in the initial recovery effort to review "documents, maps and photographs depicting conditions immediately after September 11, 2001; the various stages of debris removal and roadway restoration; and the WTC site and infrastructure in the vicinity as it appears today." This group then "engaged in lengthy and detailed discussions with those who supervised the recovery work and had firsthand knowledge of the facts and circumstances."[59] The goal of these conversations was to determine which areas were so badly impacted during the collapse of the towers that complete removal of all crushed debris was unlikely, and therefore, that a renewed search was justified. They also sought to determine which structures and infrastructure had not been sufficiently searched since the attacks.

The authors outlined the areas that the group believed to be fully cleared of human remains: West Street, Liberty Street, Vesey Street, and the World Trade Center bathtub. They then listed areas that could still potentially contain human remains. The most notable of these areas was the West Street Haul Road, where the original manhole that contained human remains was located. The report noted that when construction crews were excavating this area to provide access to Con Ed manholes they found a layer of World Trade Center debris in what was supposed to be clean fill. While the report suggested that this debris could have been the result of poor excavation during the process of the haul road construction in early 2002, the activist families believed this find indicated World Trade Center debris had been used to construct the road, which would mean that not all debris from the site was properly removed and taken to Fresh Kills for processing. The group also noted that 140 Liberty Street, which became a main entry point into the bathtub, could also contain unexcavated remains below ground. Moving out from the site itself, the group noted that manholes and utility boxes under surrounding streets could conceivably contain remains, as well as buildings and rooftops within the "vicinity" of the World Trade Center.

The report then makes several recommendations, including:

- the excavation of the West Street haul road outward from the point where World Trade Center debris was discovered until such material is no longer present;
- exploratory excavations (with careful sifting of debris) at regular intervals along the haul road and at 140 Liberty Street to ensure that no other pockets of World Trade Center debris exist;
- inspection of all subterranean structures within a zone expanding out approximately one full city block in all directions from the World Trade Center site that had not been rebuilt or routinely used;
- in the event of the discovery of World Trade Center material, the additional searches radiating out in concentric circles should be performed from that spot until no more debris was found;
- thorough inspection of Fiterman Hall, 130 Cedar Street, and the interior of the Deutsche Bank Building take place before demolition
- inspection of the rooftops of the Millennium Hotel (20 Dey Street) and 1 Liberty Plaza be searched since there was no definitive evidence that this had ever taken place.

Finally, the report recommended that a protocol be established for all future construction work that would take place in and around the World Trade Center site.[60]

The Bloomberg administration accepted these findings, and immediately began to implement them. It promised full coordination with the Port

Authority and no construction delays, which explains why the haul road, which was needed to move construction equipment in and out of the site, was not fully excavated. They estimated that the full search would take about a year.[61]

Activist family groups immediately criticized the report for being too limited in scope. They argued that a legal suit should be filed against the contractor or contractors that allowed debris from Ground Zero to be covered up during the construction of the haul road. They also noted that many areas that had previously been searched had yielded caches of remains. Thus, to search only those areas that had not been fully searched since 9/11 or were not currently under construction was insufficient. They also challenged the extent to which the sewer system would be searched (and questioned whether material was removed from the sewers in the months after Sept. 11 without being properly searched), lamented that the city was "totally disregarding the effect of the blast dynamics" in determining how far beyond the site the search should extend, and wondered why construction and utility workers were still in charge of excavations when an organization like JPAC could provide trained experts to do this work. In a press release, Put It Above Ground mocked the mayor's statements that the areas that were not going to be searched were free of remains.[62]

In a subsequent release, Put It Above Ground also claimed that the tests pits to be dug in the haul road and 140 Liberty Street were too small and too shallow and only covered a small percentage of the total relevant area: "It appears that the LMCCC is just putting on a show to make it appear that there is a comprehensive search going on. It will get good media coverage and mislead the general public. All the while construction will continue and more remains will be lost forever. It's like playing Russian Roulette with someone who knows where the blank chambers are."[63] Activist families had simply lost all faith in the city administration.

On November 2, 2006, activist family groups staged a rally to demand JPAC oversight of the recovery effort. According to news reports, at least 200 people attended, including several congressmen and state representatives.[64] In the days that followed, families continued to engage the media with their demands, and in many cases linked the discovery of new remains to their ongoing concerns that numerous human remains lay buried at Fresh Kills. Bloomberg and his deputies continued to insist that the city was doing a good job in its latest search efforts and that no new searches would take place at Fresh Kills.

While the families may have garnered support from federal and state representatives, they had less luck at the local level. In addition to the lack of success in the mayor's office, on November 15, 2006, Manhattan Community Board #1, which served lower Manhattan, considered a resolution brought by the Bloomberg administration to support the city's plan to search for remains and against the participation of JPAC. While acceptance of this resolution would have no policy impact—community boards act in an advisory role to New York government on neighborhood issues but have no actual authority—it would have demonstrated that residents of lower Manhattan supported the administration's actions. The resolution passed the executive board eight to four but was tabled until December 19, in response to calls from local residents and some board members for additional information. After a heated debate that evening, a decision on the proposal was once again postponed, this time until the January 2007 meeting.[65]

New Searches

Despite criticism from some families, the OCME and the city went about preparing for a renewed search for human remains, set to begin in December 2006 and be complete by the summer of 2007. As outlined in a November 15 memo to 9/11 families, they set up a dedicated sifting facility at 11 Water Street in Brooklyn that would process all material excavated around the World Trade Center site. Sifting of roof ballast would be done on site. They also put protocols in place for testing soil and debris for toxic contaminants, what to do in the event that remains were discovered, and where to store material once it had been sifted (Fresh Kills). The OCME and the city also arranged for members of the NYPD and FDNY to observe the operation at all times.[66]

Activist families responded immediately to the city's plans, criticizing the same elements of the plan that they had expressed concern about previously and questioning the extent to which the city was operating in good faith. "Can anyone continue to believe," they asked, "that this city has the slightest care for the families if they are considering once again placing anything from the site at Fresh Kills? This is in itself a slap in the face of all family members." By this point, the activist families had come to the conclusion that the city was hoping not to find any human remains in its test excavations so that construction could continue undisturbed.[67]

The next day, the families issued a release that presented their concerns about the latest human remains discoveries, beginning with a quotation sometimes attributed to nineteenth-century philosopher Arthur Schopenhauer: "All truth passes through three stages. First, it is ridiculed. Second, it is violently opposed. Third, it is accepted as being self-evident." They described the heartache of the more than 1,150 families who were still waiting for news of the remains of their loved ones so that "maybe just maybe they might have something of their loved ones to bury." And then they spoke of what they perceived to be the pathological lies and PR spin of the city and their desire to do whatever it took to keep construction moving forward. "So once again," they concluded, "we are left to tell the story. They may have broken our hearts but no matter how hard they try they will not break our spirit. We will continue to ask the real questions. . . . It is up to us to continue to make sure that the public knows what is really going on. We must not allow them to break our spirit. We owe it to our loved ones."[68]

A month after the city's plan to search for additional remains was made public, Diane Horning expressed frustration to her supporters about the tremendous difficulty she was having contacting the people charged with assisting 9/11 families and interacting with the public.[69] Although city personnel certainly had a good excuse for being difficult to reach—they were involved in responding to public outcry over the November 25 shooting of a young African American man by NYPD officers—Horning argued that this was exactly when the city ought to call in experts who were dedicated solely to the recovery of human remains and not running a city with the complexity of New York. As the holiday season approached, activist families kept their supporters abreast of efforts to communicate with city representatives. They detailed their difficulty getting anyone to provide more than rhetorical support of the campaign to bring in JPAC to do the recovery work.[70] They also complained that the city had been attempting to portray the families as "ungrateful crybabies . . . who will not be satisfied."[71] This time, they were discussing the mayor's announcement of the compromise concerning the way in which the names of the victims would be listed at the memorial, just as pressure was mounting on him to reconsider his decision not to call in JPAC. When the city would rather not address an issue, the families argued, officials found "something else to divert the public's attention. It's like playing three-card monte, if we play we are destined to lose."[72]

Horning's frustrations were shared by many other family members, who continued to fight for a better recovery effort at Ground Zero. In a letter to newly elected New York governor Eliot Spitzer, Jane Pollicino, wife of 9/11 victim and Cantor Fitzgerald vice president Steve Pollicino, and a prominent 9/11 activist, wrote, "In losing a loved one on September 11th I could never have prepared myself for the issues that continue to surprise and haunt my family and should have the same appalling effect on the average American citizen. In spite of the bizarre issues, probably the most frustrating part of the 'treadmill' has been how I, as a '9/11 widow,' now have a label and have been misunderstood and misrepresented." She asked Spitzer to help get the World Trade Center back on track, stop the misinformation that emanated from the agencies responsible for rebuilding, ensure that construction would not take place at the expense of uncovering additional human remains, and prioritize the needs of 9/11 families. "At the risk of sounding like an annoying 9/11 family member and living up to the image that has been portrayed, I will continue to express the distress that I feel as a result of the major missed steps that have been taken starting with the first days and weeks after the attack. Issues such as the search and recovery of the victims, the mishandling of old and new remains, the disregard of building codes, the quality of the air [in lower Manhattan] cannot be ignored."[73]

Work dragged on at the site throughout late November 2006 and into December 2006 as the city struggled to get the 11 Water Street facility up and running. Reports circulated that the Port Authority was not cooperating with the search effort, but families noted with a mix of surprise and gratitude that "there is a representative from the Medical Examiner's Office who is giving the workers a really hard time as he insists that everything is done correctly. At least someone has our backs."[74] Excavations had come to a standstill, because the city did not want to proceed until it had someplace to sift the material. The only development that occurred during this time was a shift in strategy with regard to the haul road: rather than digging a series of test pits, they would dig a five-foot wide trench (of varying depths depending on conditions) the length of the road on the recommendation of OCME anthropologists.[75] This approach made it more likely to find areas containing remains of victims and was in line with the approach taken by professional organizations like JPAC.

Finally, on December 18, the city formally opened the 11 Water Street facility and immediately announced that it would offer tours beginning in January to any family member who wanted to witness the remains recovery process for themselves.[76] With this final piece of the process in place, the city could begin to examine material from lower Manhattan. On December 29, Deputy Mayor Skyler sent Mayor Bloomberg an official update on the search for additional remains. He noted that few additional human remains were located in the seven additional Con Ed manholes and two service boxes searched in the haul road. He also gave updates on plans to search rooftops and buildings later in the fall. Most significantly for families, Skyler revealed that the trench dug along the haul road showed potential World Trade Center–related debris (including computer parts, carpet, electrical wires and steel), necessitating a further expansion of excavation efforts to include approximately three quarters of the length of the road. Skyler also reported that "more material than was originally anticipated" would be removed from subterranean structures for sifting at 11 Water Street, based on the Con Ed and Verizon survey of 430 such locations. He also indicated that the contents of approximately 165 sewer structures would be examined. Skyler confirmed that all sifted debris would be brought to Fresh Kills and stored in temporary containers until World Trade Center Families for Proper Burial's lawsuit against the city was resolved. Finally, he told the mayor that the current estimate of the cost of this latest round of human remains recovery was $30 million.[77]

Haul Road Recoveries

The first report of remains recovered from the haul road came on December 29, 2006, when OCME forensic anthropologists confirmed that it had found four human remains within the area containing suspected World Trade Center debris.[78] Then, on January 2, another two bones were uncovered from the same area, and then later in the week another nine remains were found. Thirty-nine more remains were recovered from the Haul Road were recovered on January 11.[79]

Family members and their allies who took the tour of the 11 Water Street on January 9 reported were generally pleased with the quality of the work being done by the OCME and praised the city for making the tour possible.[80] They detailed how remains were being processed (according to their understanding) and enumerated the demands that they placed on OCME officials, mostly pertaining to sharing protocols and procedures, regular reporting of results, and debunking misinformation. They also reported that the OCME stated on the record that it would fully excavate the bulk of the haul road and the 140 Liberty Street parcel. The families complained that the OCME would not provide answers to the questions they had posed months before, but noted that representatives promised to do so at a later date.[81] As of January 26, at least, families still had not received what they considered to be satisfactory answers and Sally Regenhard wrote directly to Deputy Mayor Skyler, Chief Medical Examiner Chuck Hirsch, and others to reiterate their demand for information.[82]

Families scored something of a victory a few days later when Community Board #1 passed a resolution (twenty-three to eleven with five abstentions) in favor of bringing JPAC in to manage the human remains recovery effort and soundly defeated a motion brought by the Mayor's office to reject assistance from JPAC.[83] The decision had no impact on policy, of course, but dissenting families felt that the local community was supporting their demands.

Although the city officials argued that family efforts to call in JPAC were "emotionally driven and are a major way of getting back at the City," critics on the board argued that "resistance by [the city and the mayor] generates mistrust" among families.[84] Board member Marc Ameruso noted that the search had been declared finished twice and a second set of eyes "not wrapped up in politics" might be useful, while board member Julie Nadel questioned the sincerity the mayor's office. "Why are you here?" she asked. "Why does the mayor care what the community board thinks?" Skyler, who was present at the meeting to argue the administration's case, looked "hurt" according to reporter for *Downtown Express*. Skyler reasoned, "in government, I think you have an obligation to meet with the community and answer their questions. . . . I'm going to keep coming [to these meetings] until you throw tomatoes at me."[85]

Throughout January, additional finds continued to occur every few days as World Trade Center material was uncovered in the haul road and brought to 11 Water Street for sifting. By the end of January, more than eighty remains had been recovered from the haul road, bringing the total number of remains found since 2005 to 1,063.[86] The city continued to explain that finds were the result of improper excavation and accidental mishandling during the process of loading debris onto trucks during the recovery effort. They never fully addressed the activist families' contention that the haul road was actually constructed with debris from the World Trade Center site.

The discovery of two large steel beams lying horizontally across the haul road in late September (each approximately eighteen feet long by four feet wide by two feet deep), and then another horizontal beam a few weeks later, seemed to back up the activist families' theory. After all, the only other reasonable explanation for the presence of steel beams in the haul road would have been impalement after falling from the sky. Any beams that were sent shooting downward as the towers collapsed, however, would have landed vertically in the haul road and traveled downward after piercing the surface of the ground.[87]

Activist families believed that the city's claim that the haul road had been constructed from clean fill was a lie. They also felt that the Department of Design and Construction either knew, or should have known, about this deviation from accepted protocol. Any remaining trust that activist families may have had in the integrity of city officials was disappearing quickly. Adding insult to injury, families learned several weeks after their tour of 11 Water Street on January 9 that officials had failed to inform them about one of the steps in the remains recovery process (debris had been washed and prescreened at another site in Brooklyn before it was brought to the Water Street facility). More than ever, families felt lied to, misinformed, and left out of the decision-making process. Their anger was building quickly.[88]

"How many more ways can the city invent to twist the dagger that is still lodged in the heart of every family member?" James McCaffrey asked. "A basic tenet of civilized societies is that one does not desecrate either battlefields or a burial ground. Ground Zero happens to be both."[89] In a subsequent letter to the editor of an unnamed newspaper that was never published, McCaffrey justified the actions of activist families. "The powerbrokers overseeing this shameful effort don't want the public to know that human remains continue to be found on an almost daily basis at GZ. They say the 9/11 families do nothing but scream and complain. But the families and other supporters are but a vehicle for the screams of the 9/11 dead. They scream from their grave for proper recovery and memorial. . . . These screams will haunt those who deny they exist and they cannot be silenced until those same people allow them to rest. Mayor Bloomberg, when will you listen to the wail of the forgotten?"[90]

The continued discovery of additional remains from the haul road throughout late winter and spring did little to assuage their anger. "Unfortunately we no longer seem surprised," said a Put It Above Ground member in one of their frequent human remains updates.[91] They also were not surprised when the city revealed that it had expanded its search to areas under State Highway 9A (West Street) that served as access ramps to the World Trade Center Complex and may have contained debris that had never been examined.[92] In a June 1, 2007 memo to Mayor Bloomberg, Deputy Mayor Skyler informed him that this additional effort would increase the total cost of the recovery effort by $2 million.[93]

Rather than give the city the benefit of the doubt and praise its efforts to do what it took to recover remains, the families interpreted announcements like this as evidence of the city's incompetence and unwillingness to develop a comprehensive search plan. Whereas the city was committed to letting "the forensic evidence determine which areas are searched" (i.e., dig test pits and trenches and expand out from where debris is discovered),[94] families believed that this was a hit-or-miss approach that guaranteed some remains would be left behind. They demanded a comprehensive plan in which all areas that were affected by the collapse of the towers and the dust cloud that ensued, would be carefully screened.

By June 1, 2007, when the excavation of the haul road and sifting of all material was complete, the total remains found stood at 397.[95] The start of the search for remains at 140 Liberty Street in late March brought new finds as well. By the time this effort was completed in the summer of 2007, more than 220 additional remains had been recovered from this area.[96] Deputy Mayor Skyler reported numerous other finds in areas that were just beginning to be searched, including two remains on top of 130 Cedar St. noticed during a preliminary visual inspection, eight remains in a sewer line, and two remains under West Street. The recovery of nineteen additional bones from "a ledge on an upper floor" of the Deutsche Bank Building on March 26, 2007, nearly a year after the original search there, bringing the total to 785, was simply more evidence that the city could not be trusted with the recovery effort.[97] Why had these remains not be recovered during an earlier search, the families asked. For Diane Horning, the pain and frustration was almost too much to bear. In a letter to her supporters on the day that Skyler's update to Bloomberg was made public, she wrote, "I don't think in five years since my son was murdered and thrown away I have cursed anyone, but today, I can easily say that I hope the spirits of the dead haunt these vile people for the rest of their days."[98]

Horning's efforts to communicate with the city continued for months. She demanded answers to questions about the original search, the new search, and how the remains that were discovered in the new search were being processed. Although the city provided answers to some basic factual questions, they asked for additional time to answer those questions that addressed issues of protocol and procedure.[99] From correspondence between the families and the city, it is clear that the families believed they had a fundamental right to participate in the decision-making process and that the city had an obligation to turn over all information related to the search. The city clearly did not agree.[100]

While one can question the families' views on what their role ought to have been in the recovery process, their consistent claims that there was far more World Trade Center debris still left buried at the site than the OCME and the city initially expected were vindicated. Although the city had originally planned to conclude the new recovery effort in the fall of 2007, by July Deputy Mayor Skyler had to admit that "our experience over the last nine months and the ongoing rebuilding of the World Trade Center site and surrounding area suggest that the search operations will continue in one form or another for the foreseeable future. We will add additional areas to the search plan as necessary, and OCME will likely have a presence at the WTC site for some time to come."[101] This statement came in response to the discovery of additional World Trade Center debris on the perimeter of the site, including across the street at World Financial Center by Port Authority contractors digging a tunnel to connect the two complexes.

By July 3, Skyler reported that 1,462 new remains had been uncovered since the search was renewed in 2005. Although the majority of these remains were tiny bone fragments, some were quite large and potentially identifiable. Families, of course, continued to press the OCME and the city to install trained professionals to search the entire area around the World Trade Center, not just areas where remains happened to be uncovered by construction crews, which they dubbed the "look what we found" approach to disaster archaeology. The accidental find in the Con Ed manhole "has become the blueprint for the hit and miss plan of the current search," they lamented.[102]

The recovery of remains around the borders of the World Trade Center site continued throughout the summer and fall of 2007—1,772 as of December 10, 2007.[103] The OCME and the city sent daily updates about the recovery effort and communicated directly with these families to reassure them that they had everything under control and that all relevant areas were being searched (i.e., those in which potential World Trade Center debris was uncovered). They also told families that the entire process was being managed by highly skilled OCME anthropologists and archeologists with "advanced degrees in their respective fields" and extensive knowledge of soil analysis, soil stratification, and excavation techniques.[104]

The Deutsche Bank Building continued to be a source of intense embarrassment and frustration for all involved in the redevelopment of the World Trade Center site.[105] Whether it was wrangling over environmental safety at the site, the death of two firefighters in a blaze caused by construction workers smoking on the job in August 2007, or allegations that the company subcontracted to take the building down, John Galt Corporation (named after a central character in Ayn Rand's novel *Atlas Shrugged*), was a shell company with a sketchy history and potential mob connections, the building quite literally would not go away.[106] Indeed, the demolition process, which was painfully slow because it had to be done floor by floor with intense government oversight, only reached ground level in February 2011—nine-and-a-half years after the attacks of September 11.

On December 11, 2007, the city announced that it was entering a new phase in the recovery effort. After $38 million spent excavating and searching through 15,000 cubic yards of soil and debris, and hand sifting nearly half of that material, the city announced that it was closing the 11 Water Street facility and "converting the operation to a mobile, forensic recovery unit that will be deployable as needed." This move was described as a shift in approach, and the city explicitly stated that "at no point in the near future would it be prudent to declare this search 'over.'"[107] Indeed, the continued need for sifting equipment, whether stationary or mobile "acknowledges the sad reality that the search for remains will go on as long as there is excavation activity in and around the WTC site."[108] By December 10, 2007, four new victim identifications had been made from remains recovered from the Deutsche Bank Building and three new identifications had been made from the haul road remains. The OCME continued to collect dirt, debris, and roof ballast, and recovered seventy-two additional remains when it sifted this material between April 5 and June 18, 2010.

All told, as of June 2010, 1,845 remains had been recovered since 2005.[109] Sifting operations were ramped up once again from April to June 2013 to examine dirt and debris collected over the previous three years. This effort netted approximately eight-nine additional remains, bringing the total to 1,934.[110]

8

Who Owns the Dead?

On an unusually foggy Saturday morning in May 2014, a column of emergency vehicles made the short journey from the Office of Chief Medical Examiner to the National September 11 Memorial and Museum to deliver nearly 8,000 unclaimed and unidentified remains of victims of the World Trade Center attacks to their final resting place: a repository behind a wall of the soon-to-be-opened museum. The solemn procession involved members of the FDNY, NYPD, Port Authority Police, and others who had participated in the recovery and rebuilding effort. Many family members of the victims also attended—some to commemorate the bittersweet return of the remains to the site, and a few to protest it. For those who approved of the plan to inter the remains within the walls of the museum, this was a moment of relief. The remains had been in temporary storage for the past twelve-and-a-half years, and they were finally being laid to rest in the place where they belonged—at bedrock within the bathtub of the World Trade Center site. For the dozen or so relatives who were present to protest the decision, on the other hand, the bones of their loved ones were about to become curiosities to attract tourists willing to pay the $24 entry fee to the museum and perhaps purchase a souvenir at the gift shop on the way out.

These dissenting families made their case to the public by wearing black gags, symbolizing their view that the powers-that-be had systematically ignored their requests for an above-ground tomb for the World Trade Center remains. In addition to the gags, they held up signs with slogans like "No Voice, No Choice" and "Human Remains Don't Belong in Museums."

One sign implied that Osama Bin Laden's remains were treated more respectfully when they were buried at sea than the remains of the victims of the attacks he had ordered. Perhaps the sign that most directly captured the views of these protesters was the one held by Rosaleen Tallon: "We are their families. . . . they [the victims] belong to us. Not a museum, not the city!!!"[1]

After Sciame

With the Sciame Report in hand in July 2006, the Lower Manhattan Development Corporation and the World Trade Center Memorial Foundation had their marching orders. Given a budget of approximately $500,000,000, they were to build three facilities: the World Trade Center Memorial on the plaza at street level; a memorial museum below ground; and a visitor orientation and education center between the two.

While the Sciame Report had laid out a new plan for the memorial and a basic structure for the museum, it did not address the repository for unclaimed and unidentified remains. Nor did it give any hints about how to balance the new dual roles of the memorial museum: to honor the dead, and to educate the living about how they were killed. That task would fall to the museum leadership team, headed by Alice Greenwald, who was hired in April 2006 after nearly twenty years at the U.S. Holocaust Museum.

Although Greenwald had extensive experience dealing with issues of collective memory and public history in relation to contentious and tragic events, she rarely had to deal with the long-term storage of human remains at the Holocaust Museum. When Greenwald and her team were brought onto the project, it was understood that the museum and memorial would be separate entities. Furthermore, Greenwald was under the impression that the storage of human remains at the World Trade Center site was a settled matter.[2] Even though the memorial and museum had been merged, she and others at the LMDC and the Memorial Foundation did not think the plan either had to, or ought to, be revisited. She and her colleagues stated that it was the will of the majority of families (or at least the groups that represented them), that the remains be stored at bedrock in a secure facility, accessible by family members and the OCME only. This, after all, had been the request of the Coalition of 9/11 Families from 2002 to 2006, and the feedback given to Frank Sciame during the redesign process. Thus, the efforts of a small group of families—many of the same ones who opposed other aspects of the way the redevelopment of the site was taking place—to bring the human remains repository above ground beginning in 2009 took museum staff by surprise.

In order to understand this controversy, it is necessary to return to 2002, when, as a result of the requests of many victims' families and family groups, the LMDC made a commitment to "create a permanent location for the unidentified remains at the World Trade Center (WTC) site."[3] The Families Advisory Council (FAC), which was created to gather input from victims' immediate relatives during the planning process, validated this plan in their draft memorial recommendations in 2002. The LMDC informed victims' families by mail between 2002 and 2006 about plans for the memorial and museum quadrant, and received feedback from them along the way. The LMDC also sponsored public forums and sought to maintain an open and direct dialogue with families and family groups.[4]

Clearly, National September 11 Memorial and Museum officials believed that they and the LMDC had gone to great lengths to solicit the views of family members and their representatives and that there was a strong consensus among families that the remains ought to be stored at bedrock. They even noted that the LMDC had made many changes to the memorial and museum based on the families' strong views, particularly on the names issue and the symbolic vessel for remains.[5] For memorial and museum officials, the changes wrought by the Sciame plan did not fundamentally affect the repository. Even though it would no longer be located within the memorial or "distinct from other memorial structures like a museum or visitors center," as described in the 2003 competition guidelines, it would still be placed within the slurry wall at bedrock, between the tower footprints, as the most vocal and active families seemed to want.[6]

Under the new plan, access to the repository would be administered by the memorial museum and the two spaces would share a wall. Memorial and museum staff argued that this was a separate space, not technically a part of the museum. Although a message would acknowledge the human remains behind the wall, the public would have no access. They would be reminded of the human loss at the site, and would be able to pay their respects to the dead, but there would be no risk of visitors gawking at remains or disrupting families seeking to be close to their loved ones. Families would have a separate space behind the wall out of the view of ordinary visitors.[7]

Families and family groups did not express any concerns about the change in plans after the Sciame Report, partly because the plans were not yet clearly laid out and partly because memorial staff carefully managed the museum design process. The staff encouraged small, carefully controlled "conversations" from the summer of 2006 through 2008, which "were designed to inform the program and exhibition development of the Memorial Museum."[8] In their summary of this conversation series, the staff noted that "prominent thinkers on museums, memorialization, American history, and the continuing impact of collective trauma, as well as representatives of key constituencies, were invited to speak with staff and stakeholders in the Museum development process. These strategic conversations were intended to help crystallize the working assumptions that will inform the planning process, and help the team develop a shared vision and vocabulary for what the Memorial Museum could become."[9]

The public report from this "conversation series" noted that "the group has been small enough to ensure meaningful conversation and large enough to engage a variety of opinions. In fact, lively debate has been welcomed as a key component of the creative process."[10] According to the public report, the conversation group included "family members of 9/11 victims, survivors, appointed liaisons from the NYPD, FDNY and PAPD, local residents, downtown businesses, interfaith clergy, architects and landmark preservationists, colleagues in the museum and cultural communities, government personnel, exhibition designers, and other interested parties."[11] The list of participants, which included Beverly Eckert, Mary Fetchet, Marian Fontana, Anthony Gardner, Lee Ielpi, Edie Lutnick, Nikki Stern, and Charles Wolf, suggests that the museum staff chose some family members that it knew shared their vision, as well as others that had been critical of redevelopment efforts but had shown a willingness to compromise. Notably, no families active in the human remains issue were invited to participate in the conversation series (although Diane Horning was invited to offer her views on the display of artifacts that may contain human remains as a guest speaker), nor were any families that had been involved in legal action against the city or its agencies. Glenn Corbett described these events to me as "magical sessions" in which certain people were carefully selected to represent all family members without actually being elected or deputized by them to serve this function. Regenhard went even further, describing the people who participated in the conversation series as "handpicked rubber stampers."[12]

The first conversations with museum, memorialization, and history experts in the summer of 2006 focused on five key issues: "The potential and character of twenty-first-century museums; the particular requirements and sensitivities of memorial museums that must balance the concerns of privacy with the imperative to educate; the challenge of understanding the Museum's role and responsibility to present visually-difficult imagery without re-traumatizing the public and in an age-appropriate way to younger visitors; how the story told in the Museum will contribute to the writing of history, and how the emerging story already echoes key themes of the American historical narrative; and the role museums increasingly play as instruments of civic renewal."[13]

While human remains were not a core subject of this first round of conversations, Greenwald discussed the issue briefly on July 20, 2006 when she introduced conversation group participants to Edward Linenthal, a historian who specializes in memorialization of war and other tragic events. She told them that she first met Linenthal during an intense period of debate at the United States Holocaust Memorial Museum (USHMM) about whether or not to put the hair of victims on display, as was done at the Auschwitz-Birkenau Memorial and Museum in Poland. The purpose of the Auschwitz exhibit was to show that the Nazis had industrialized mass murder to the point that they had commoditized its human byproducts. The Nazis were selling human hair to various industries to be turned into finished products such as yarn, felt, stockings, and mattress stuffing in order to fuel the war effort. Some people involved in the planning of the USHMM argued that hair was necessary to tell the full story of the Holocaust, while others argued that displaying hair would be contributing to its continued commodification since it would have to be obtained from German institutions. Doing so was disrespectful to victims and survivors. The debate was put to rest, however, when an advisor to the planning committee, who was herself a Holocaust survivor, stood up at a meeting and pointed out that her own mother's hair could possibly be put on display and this fact made her uncomfortable.[14] This intervention reminded the committee members that the hair to be displayed "was human 'matter' out of place, registering differently from railcars or shoes" and thus requiring a radically different level of sensibility and thought from most other artifacts of the Holocaust.[15] The episode suggests that Greenwald was attuned to the sensitivity of human remains for victims' relatives.

By the time the conversation series resumed in the fall of 2007, the National September 11 Memorial and Museum had selected Thinc Design (along with a company called Local Projects) to serve as the primary exhibit designer. The museum planning conversation series event on November 19, 2007 focused directly on the issue of human remains at the museum, both in the context of the repository and in the display of artifacts. Summarizing the conversation on the repository, memorial and museum staff wrote that "overall, the comments reflected a favorable view of the proposed placement of the OCME office in a prominent location within the East Chamber at bedrock, behind a great wall running along the eastern perimeter of the Museum in the space between the two tower footprints." This, according to one participant, was the symbolically "charged center of the site." The participants in the outreach efforts also supported acknowledging the existence of the remains for museum visitors.[16]

Thus, the museum staff believed that the families of victims were on board with the idea that the repository should be part of the museum experience and that it would in fact be immoral and ahistorical to hide the presence of remains at the site. Initially, activist families did not speak out about the location of the repository, primarily because plans had not yet been completed and made public. According to activists, it was not until April 2009 that they fully understood the nature of the repository design and its position within the museum.

The Composite

As the conversations series summary noted, the museum had to balance sensitivity toward family members with their commitment to "factuality and authenticity in telling the story" of September 11. This is, of course, a dilemma for all memorial museums, and one that has no settled solution. Many museums displayed human remains as seemingly indisputable evidence that the tragedy in question had taken place. Human hair at Auschwitz, damaged skin at Hiroshima, and skeletal remains in Cambodia and in Rwanda make it harder to deny the events being memorialized. While museum planners had no intention of putting human remains on display, they wanted to demonstrate the extremity of violence produced by the collapse of the towers. One way they hoped to do this was to display an artifact that had been dubbed the "Composite," which was the product of intense heat and pressure compressing four or five floors of one of the towers into an object roughly four feet in height. The Composite was particularly impressive because, "within the moonscape of its surface, intimate evidence of human activity is visible: bits of paper are present, the imprint of writing still discernable, preserved through carbonization."[17] Shockingly, even though paper had been preserved within the Composite, the OCME had come to the conclusion that this object and others like it did not contain any human remains because the temperatures necessary to create it would have completely consumed any biological material present.

Immediately on reading a news article about plans to display the composite, Diane Horning wrote to Greenwald.[18] Rather than conceiving of the Composite as a dramatic and historically authentic tool to convey the force of the collapse of the towers, Horning argued that it was a "tomb of trapped remains" that should be buried in a dignified manner if they were truly unidentifiable.[19] Further, she questioned the OCME's contention that any remains in the Composite were beyond recognition. In a follow-up e-mail to Greenwald, she wrote: "You ask us to believe that in this one relic, paper has survived, but human remains are too 'seared for identification.' This disparity defies reason. It is possible that the dissection of these composites might actually yield identifiable remains." Linking the issue of the display of the Composite to her other activism, Horning continued:

> The City has left our dead in the garbage dump and in sewers and rooftops throughout the WTC area. Now you want to take some few remains we actually have and deny them a proper burial as well. Unless you know exactly who is crushed inside this "composite" and can get permission from the surviving families to exhibit these remains, you simply cannot desecrate these dead any further. Why you feel the need to do such a thing is beyond me and speaks to a great lack of compassion. It is not as though there are not myriad other relics that can convey the force and power of that day's destruction. Why must you use a "relic" that contains human remains? I know that the memorial foundation is very fond of pointing out that this is NOT the families' memorial (or museum) and we are only one of many "stakeholders." But, when it comes to the dead, there can be no question of who owns the dead . . . and that is not you or the entities you defer to. It is the surviving families. The plan to display the dead is an obscenity and an indignity that cannot be tolerated.[20]

Greenwald responded to Horning that she appreciated her concern and reassured her that the museum took great care in deliberating on the inclusion of the Composite in its displays. She noted that she understood that the potential that human remains could be found in the Composite meant that the artifact had to be treated with "an exceptional level of reverence and respect, both in its placement and its preservation." She went on to note that the World Trade Center Memorial Foundation's Program Committee (which was comprised of family members Howard W. Lutnick, Paula Grant Berry, Debra Burlingame, Monika Iken, Anthoula Katsimatides, and Tom Rogér, along with actor Robert De Niro, Emily K. Rafferty, the president of the Metropolitan Museum of Art, and Seth Waugh, the CEO of Deutsche Bank Americas) had approved of the display of the Composite "within a sensitive and meaningful environment."[21]

Bill Doyle, who had been copied on the correspondence by Horning, wrote back to Greenwald accusing her of stacking the program committee with "YES people," those who would help advance the mission of the museum without offering much resistance. He argued that the few family members on the committee were not representative of the majority, particularly on the issue of human remains.[22] For Doyle, father of twenty-five-year-old Cantor Fitzgerald employee Joseph Doyle, any decisions involving human remains ought to include those organizations that had been created specifically to address these issues.

Over the next few years, Horning, Regenhard, Doyle, and other activists tried to convince the museum to search for remains within the Composite, and short of that, not to put it on display. As late as 2012, the museum had not publically stated what it would do with it. Ultimately, the museum staff decided that the Composite was a significant artifact that helped tell the story of September 11. Recognizing the potential traumatic reaction that some people would have to the artifact, however, it was put on display in an alcove in an out-of-the-way corner of the museum so that visitors could choose whether or not to view it. Thus, while it may be a draw to the museum, it is not a "traffic stopper" and is not displayed in a sensationalistic way.

The Repository

On April 6, 2011, retired FDNY deputy fire chief Jim Riches submitted an outraged op-ed piece to the *New York Daily News* with a provocative title

"Who owns the dead?" Riches, who lost his twenty-nine-year old firefighter son Jimmy in the World Trade Center attacks, used the piece (which was never published but was widely disseminated among activist families) to communicate his disgust with the way that unidentified and unclaimed human remains found at Ground Zero would be interred at the redeveloped World Trade Center site. It was the same basic question that activist families had asked in 2004 in connection with the remains being stored at Fresh Kills. In his op-ed, Riches argued that memorial and museum staff, the LMDC, and the OCME had usurped the "sole right" of next-of-kin to determine the fate of their loved ones' remains. The decisions made by these actors—specifically about the storage of unclaimed and unidentified remains seventy feet below ground behind a wall of the museum complex rather than in a separate memorial—angered families of the dead, thwarted democratic decision making, and explicitly reneged on promises made to families and the public in the months after the attacks and during the 2003–2005 memorial and museum planning process.[23]

He explained that, in order to reach their loved ones' remains, grieving families would have to "walk past the 9/11 Museum store on [the] 1st floor, selling its WTC cups, shirts, books while on their way down past thousands of tourists." In place of this plan, which had emerged over years of discussion, debate, crisis, cost-cutting, and controversy, Riches wanted an above-ground tomb of the unknowns at street level and watched over by a twenty-four-hour honor guard—which would be a fitting tribute to the hundreds of uniformed service professionals who had lost their lives and the civilians who had died with them. "The 9/11 Memorial and Museum will be a popular site for millions of Americans and tourists every year, but NEVER FORGET it will always be a cemetery for the 9/11 families who lost so much that day." Riches closed by demanding that the museum conduct a poll of all families who had lost a loved one in the attacks to ask them whether they preferred the current proposal to store the remains seventy feet below ground behind a wall in the museum complex that could only be reached by pushing past thousands of entrance-paying tourists or in a "respectfully accessible" above-ground tomb completely separate from the museum.[24]

What Counts as Consultation?

The roots of Riches' outrage dated back nearly two years, to April 16, 2009, when Sally Regenhard and a few other family members first learned about plans for the repository when they went to hear Alice Greenwald speak at a Downtown Alliance event at St. Paul's Church. Before this talk, which was titled "Passion on all Sides: Planning a Memorial Museum at Ground Zero,"[25] Regenhard says she had no idea that there would be a connection between the repository and the museum—her understanding was that the two structures would be totally separate and there would be no physical link between them—as was the case in early iterations of the memorial and museum design. She was quite surprised that the repository would be completely underground, unlike all of the previous publically available renderings that showed it at the base of the memorial but still open to the sky (in Arad's initial design the waterfalls were not closed off, leaving the area that was to contain the names and the repository open).

Regenhard was shocked and informed other activists about what she learned. They were granted two meetings with museum staff during the summer of 2009 to learn more about the museum's plan and to express their concerns. Regenhard and her advisor, Glenn Corbett, also talked to David Hurst Thomas, a curator of anthropology at the American Museum of Natural History in New York,[26] about the obligations of museums in the context of human remains and whether the National 9/11 Museum and Memorial had engaged in appropriate levels of consultation when deciding where to place the repository. Thomas was well-known for his research and writing on Native American archaeology and had written a popular book on the ownership and control of native skeletal specimens entitled *Skull Wars: Kennewick Man, Archaeology, and the Battle for Native American Identity*. He was sympathetic to the cause of the dissenting families and asked a colleague, Chip Colwell-Chanthaphonh, curator of anthropology at the Denver Museum of Nature and Science, to advise them on museum ethics and practice. Colwell-Chanthaphonh had extensive experience dealing with Native American remains and sacred objects as the museum's officer responsible for carrying out the museum's obligations under the 1990 Native American Graves Protection and Repatriation Act (NAGPRA). This act requires museums to engage in dialogue with lineal descendants, Indian tribes, and native Hawaiian groups about repatriation of the remains of their ancestors, as well as certain culturally and religiously significant objects collected by the museum for scientific purposes or public display. These items were taken during a time in history when archeologists, anthropologists, art dealers, private collectors, and the museum community did not seriously think about the effect that such removals would have on native communities.

At the heart of NAGPRA is the right of lineal descendants to determine the fate of remains and objects in museum collections.[27] Deciding whether to repatriate or continue to display or store remains and objects should be collaborative, according to NAGPRA, and the final say always rests with these descendants. Museum officials cannot overrule or ignore the desires of lineal descendants, Indian tribes, and Hawaiian groups. There have, or course, been numerous debates about who gets to speak for native groups, as well as the necessity, burden, and effects of NAGPRA. But the bottom line is that the rights of lineal descendants to determine the fate of remains of their ancestors is now well-established by law.[28]

Colwell-Chanthaphonh called consultation with lineal descendants "the duty of ethics that exceeds the dictates of law."[29] Further, he stated that the American Association of Museums' Code of Ethics affirms that the " 'unique and special nature of human remains . . . is recognized as the basis of all decisions concerning such collections, and directs that 'competing claims of ownership that may be asserted in connections with objects in its custody should be handled openly, seriously, responsively, and with the respect for the dignity of all parties involved.' "[30]

After consulting with Thomas, Corbett, Regenhard, and other family members in 2009 and early 2010, Colwell-Chanthaphonh came to believe that there were serious problems with the way the National September 11 Memorial and Museum was engaging with victims' families and that it needed to alter its consultation practices. In an *Anthropology Today* article published in 2011, he summarized his views: "The 9/11 Museum has never systematically asked all family stakeholders for their input on its decisions. While it has made attempts to include some families, these efforts have been tightly controlled. Many outreach events are by invitation only, and many families have never been invited."[31] He further argued that the museum did not have a clear and consistent definition of who counted as a relevant family stakeholder. He pointed out that when Diane Horning was invited to speak at a conversation series event against the display of the composite in the museum, museum staff used testimony by Debra Burlingame, whose brother Chic was the pilot of the hijacked plane that crashed into the Pentagon, to rebut Horning's claims. Such a tactic was unreasonable, Colwell-Chanthaphonh argued, because Burlingame "had no direct stake in the World Trade Center remains."[32]

Finally, Colwell-Chanthaphonh indicated that, while the National September 11 Memorial and Museum had the capacity to reach out to the entire population of World Trade Center victim families (i.e., lineal descendants and/or next-of-kin), they had done so on only two occasions since 2006: once to make sure that the names of the deceased were spelled correctly and once to ask families for mementoes of the victims for displayed. They had never asked all family members how they wanted unclaimed and unidentified World Trade Center remains to be stored. "The rationale behind outreach choices is opaque," Colwell-Chanthaphonh concluded, suggesting that the National 9/11 Memorial and Museum did not actually want to know what all families thought of their plans.[33]

Concerned about what they saw as inadequate consultation by the museum, Colwell-Chanthaphonh and Thomas decided in late-winter 2010 to start a petition within the anthropology community to try to convince the museum to change. Soon after the petition was drafted, however, someone informed the administration, who implored Colwell-Chanthaphonh and Thomas to talk to them directly. They agreed, on the condition that the family members could be a part of any discussion.[34] Colwell-Chanthaphonh told me that museum officials resisted this request, but he and Thomas insisted that the families be involved in any meeting because this was the ultimate goal of their intervention.[35]

After much discussion, museum officials at first relented to the demand for a meeting involving all parties. They then rescinded the offer, stating that they had already engaged in extensive outreach and that all relevant information could be found online. Colwell-Chanthaphonh and Thomas did not find any evidence that all families had been consulted since Sciame's redesign, however, and continued to press their case. Colwell-Chanthaphonh also told me that, around this time, Alice Greenwald called the presidents of his and Thomas's museums to discuss what she claimed were their unprofessional activities. Colwell-Chanthaphonh called this "an overt threat to get us to distance ourselves from the families." The presidents of both museums were completely supportive of the two anthropologists, however.[36]

Who Are Legitimate Stakeholders?

In April 2010, the National September 11 Memorial and Museum once more agreed to meet with all stakeholders, but only at their offices. While Colwell-Chanthaphonh and Thomas suggested that the meeting should be held in a place that wasn't emotionally fraught for the families, the museum officials would not budge. The meeting finally took place on June 8, 2010 at the museum's offices at One Liberty Plaza.[37]

Two main issues dominated the discussion, and conversation wove back and forth between them. One was conceptual: would the remains be a "programmatic element" of a museum charging entry fees or would they be part of the memorial aspect of the site? The second issue was procedural: through what mechanism should the museum have decided how remains should be handled at the site? This question led to a host of others. Did officials adequately inform and consult with the relevant stakeholders on the issue of remains (i.e., the families)? Who speaks for the "average" family member? Could this role be meaningfully delegated to a small number of people without consulting the entire community? How should the silence of the vast majority of families on the issue of human remains be interpreted? Did the fact that most family members never corresponded with the LMDC or the memorial and museum staff about human remains mean that they were largely on board with the plans to inter them behind a wall of the museum? Did it mean that they did not want to even think about this issue or that they felt that their opinions would be ignored? Or did it mean that they simply were not adequately informed about the plans?

With respect to the issue of whether the remains would be part of the museum, the families and their advisors returned again and again to the October 3, 2006 request for qualifications (RFQ) for exhibition design services the museum had put out as proof that museum staff and the foundation considered the remains part of the museum experience.[38] The RFQ describes how visitors would enter the museum through the ground level visitor center and then move down to the "Memorial Hall" at thirty feet below grade, eventually arriving "at the bedrock level, some seventy feet below the Memorial Plaza, where a variety of programmatic elements are expected to be offered." The first "programmatic element" listed is the "repository for the unidentified remains of the World Trade Center victims of 9/11."[39] Other elements listed include "monumental artifacts" from the site, a narrative exhibition of the events of the day, contemplative spaces, an exhibition for families and young children, digital resources, and archaeological exhibits featuring structural remnants of the twin towers. As Colwell-Chanthaphonh argued, "the 2006 document makes plain that the human remains are to be an integral part of the museum—no longer resting in an accessible and separate space for memorialization and grieving, but operating as a museological feature."[40]

Based on their responses, museum officials seemed befuddled by the families' insistence that the museum planned to use human remains to attract visitors to the museum. They did not interpret the language of the RFQ in the same way that the families did. Indeed, when Glenn Corbett asked National September 11 Memorial and Museum director Joe Daniels "who actually decided to make the remains a 'programmatic element' of the museum?" Daniels was perplexed. "How do you define 'programmatic element'? What do you mean?" he asked. The conversation continued for a bit, before Daniels finally grasped the root of their differences. For the families, the remains themselves had become a part of the museum, but for museum staff, it was only *the wall* that separated the museum displays from the OCME facility that was a "programmatic element" in need of design.[41] Elaborating on his views later in the conversation, Daniels stated that the National September 11 Memorial and Museum "implemented what we believed was an expression of how the human remains should be dealt with. And the way we have it right now is that there's a wall—that is there is a wall of the Medical Examiner's Office, which is also a wall of the museum behind which the human remains are being kept. That's it. There's no flashing lights saying come see the human remains. And that whole suggestion is just ridiculous. All I'm saying is this is specifically how we have them, which is a demarcation of behind this wall is the unidentified remains of the victims of the World Trade Center attacks."[42] For memorial and museum staff, the repository was part of the story that was being told on the site of one of the worst massacres in American history. It was not commemorating an event far off in another place, but rather telling the story of an event that happened on the site just a few years before.

Corbett then followed up by saying, "I'm just trying to figure out—between 2003 and October of 2006 something changed. . . . The memorial sort of left and the remains stayed behind. And I'm just trying to figure out who made the decision to incorporate the remains into this museum. That's what I would like to know."[43] Daniels replied that there was such strong advocacy to return the remains to bedrock that even when the plans changed it did not affect the overarching decision to return the remains to bedrock. Corbett continued, "The reason I'm asking is because I think it was fairly clear back in 2003, 2004, that the memorial was to be separate and distinct from any museum or visitor center. And then when you get to October of 2006, now the remains are part of the museum. I mean you can say there is a wall between them, but it's still part of the museum. I mean, you're calling out those remains as part of the museum. And I'm trying to figure out who made that decision and when was it made." Joe Daniels answered:

> Well, just to be clear about that. There was never—the statement itself, and this is an important thing because it was, it was signed off on by families as well as the political stakeholders, the statement is not that the repository should be a separate space, [the statement is that] . . . the memorial itself should create a unique and powerful setting that will be distinct from other memorial structures like a museum or visitor center, [make] visible the footprints of the towers, include appropriate transitions to or within the memorial. So, there is no mandate anywhere that says the repository of human remains needs to be separate from the memorial.[44]

For families and their advisors, though, it was clearly implied in all publically available plans that the remains would be stored within the memorial, not the museum, and this is what families originally agreed to. The change in design was so significant that, in their view, it negated any previous agreements that had been made because the facts had shifted so significantly. To which Daniels replied, "So basically what you're saying is, is that should the organization have assumed that the very strong advocacy led by many people here to move the names above ground, should that have automatically included moving the remains above ground and, I mean, we can, that's a discussion we've had and I personally don't think that those, one follows from the other."[45]

In response, the families and their advisors got to the heart of their critique. They argued that the National September 11 Memorial and Museum had not adequately consulted with the lineal descendants of all individuals whose remains could potentially end up in the repository. By carefully selecting a few family members who they knew would be willing "rubber stamp" the decisions of the museum staff and its advisors, they short-circuited the consultation process. Specifically, the families advocated a strongly democratic decision-making process in which each relevant stakeholder had a single vote and majority would rule. Delegates could not represent the interests of stakeholders unless explicit permission was granted to speak in their name, and silence could not be taken as a tacit form of consent. In order for the process to be legitimate, stakeholders needed to be polled directly about their views on the issue after they had been fully informed of the situation and possible alternatives.

Rosaleen Tallon was particularly forceful on this issue. Her contention was that the memorial and museum had not done enough to inform the "average 9/11 family member" about what happened with the remains at the World Trade Center site in the wake of the Sciame redesign. In her view, the family member who "tried to get their life going, raise their children" and did not follow what had happened at the site closely knew nothing about the plan to inter the remains at bedrock as part of the museum complex. Further, even those who did know had not been given a chance to offer their views about what they thought of the plan. She said, "I would use my mom as an example for that because even though I'm so involved with the memorial, between raising the three kids and she has a house on her own now, we don't talk about this stuff. So, if I had asked her a month ago before I really got into the meat and potatoes of this with her, what she thought was happening to the remains. Now this is a woman who gets her mailings, she gets the Tribute Center mailings, and she'll read whatever she gets from the fire department. She had no idea what was happening to the remains. That's the average 9/11 family member."[46] Tallon asserted that the museum had an obligation to inform the families about changes made to the layout of the museum and memorial and to ask for their input on the situation. She disputed the memorial and museum's claim that the consultations about human remains had been adequate.

In the families' view, the memorial and museum had an obligation to inform families about the changes after the Sciame Report and to allow them to provide their feedback. "I would expect that the human remains would be equal if not more important than how our names are listed," Tallon stated. "We received an official mailing on the names," she continued. "My mom had to go to the post office to sign for it. It was registered. It was official. It was important. People felt like, 'I have to make a decision on this.' We read the pamphlet—this was how it was going to be, as they say, 'set in stone.' . . . [T]o me the human remains deserved that much of an outreach, and if not that but more, where we sat down to talk about this. As I keep reiterating, the families are clueless—and that's not a reflection on the families. That's a reflection on you. And I really truly feel that if you were to reach out to the families you would really find out that we're not just saying this to be a thorn in your foot, we're saying this because we truly feel that the input hasn't been there."[47]

Regenhard reiterated Tallon's concerns, saying that all families, not just a few, "need to have a voice and a choice."[48] Horning went even further, stating that families were the *only* stakeholders who mattered when it came to the disposition of the human remains.[49] In order to achieve this goal, the families proposed that the memorial and museum use its mailing list to inform all World Trade Center families about the plan to house remains in the repository and ask for their feedback and approval.

For Greenwald, the complexity of the World Trade Center situation made direct democracy challenging. There were so many interest groups to negotiate among and because they "couldn't fit [every family] in the room, frankly."[50] For her, the consultation process, in which the museum heard representatives of each stakeholder, was the only reasonable way to move forward. Greenwald directly acknowledged that the Sciame redesign had fundamentally altered the relationship between the memorial and museum, but noted that Sciame had worked hard to gather and consider the opinions of as many families as possible. She continued that during the Sciame redesign,

> the museum building, which was on another part of the plaza, became, frankly, vertically integrated with the memorial. This was a big change. When I took the job, I was actually going to build a different museum, okay? Because that was the design before I arrived. Within weeks of my arriving the whole project had become integrated in a way it hadn't been before. So I totally understand the confusion for anyone who's outside this process to understand fully how this evolved.[51]

Greenwald then insisted that the placement of remains at bedrock in between the two tower footprints came at the request of family groups, and that memorial and museum staff saw no problem with the repository being in the vicinity of the museum. She told families:

> I think the distinction in our minds—and clearly wasn't universally held and understood—was that the memorial in the museum, and the memorial plaza, is entirely a distinct precinct and it's an integrated precinct. It is all about memorialization. We are a memorial museum. We're not another kind of museum. We're not a natural history museum. We're not just a history museum. We are a specific kind of

> museum and they are around the world right now, and they are called—memorial museums. When you are a memorial museum, there are certain obligations of sensibility, of museum ethics, of consideration of the constituencies, understanding that there will not necessarily be unanimity of opinion, I mean that is just a given.[52]

While Greenwald admitted to families that her job would have been "a whole lot easier if [the repository] had not been part of what I was given to design around," she was adamant that putting the repository at bedrock was an obligation to families.[53] She said that the job of the staff was to work with families to ensure that their vision was preserved during the design process. "We certainly did not do it with all 3,000 family members, and that is a point," Greenwald continued, but "we've looked at this carefully. We've met with interfaith advisors to talk about how this should actually happen in the most respectful way. From our perspective—understanding that it may not be to you sufficient—we feel that we have tried to approach this with a great deal of candor, communication, involvement, listening, presentation. I mean the fact is you heard the presentation in a public forum. We're not hiding anything. We have been speaking about this since I've been here."[54]

For Greenwald, the consultation process had worked, even if a few families did not agree with its conclusions. Indeed, she believed the placement of the remains to be a settled decision, and would not consider reopening the decision-making process because of the concerns of a small number of families.[55] "We're not going to please everybody," she told the families. "I know that. In fact the thing I say all the time is that the only true thing about this project that you can guarantee is that we will make everybody unhappy about something. That's the truth. It's just the nature of something this charged, this difficult, and this recent."[56] Greenwald recognized the special importance of families, but explained that "we're trying to navigate through all of the rapids and all of the boulders and all of the difficulties to create something that, we believe, to the best of our ability as museum professionals understanding museum codes of ethics, understanding the issues that you deal with with NAGPRA. This is a slightly different situation. It is a different circumstance. 9/11 is an unprecedented event. It's not typical."[57] Greenwald concluded by arguing that, although not every family member had a chance to voice their opinion, the consultation process included representatives of the families.

Tallon challenged her on this statement: "Represent? But when you talk about—representative families and you're talking about human remains, they're talking about what they want specifically. I know that the remains are grouped together right now, but they're thinking about their loved one. They're not advocating for everybody else. They're specifically talking about what they would want for their son or daughter. So I think talking to people as a representative body for that is not fair because they're getting to have a voice whereas the rest of the people aren't."[58] Colwell-Chanthaphonh followed up, arguing that whenever possible museums are required to consult lineal descendants rather than representatives to discuss issues of human remains. He then suggested that the National September 11 Memorial and Museum send a letter to all next of kin about the disposition of remains within the repository behind the wall of the museum.

At this point, Daniels seemed to lose his cool, asking, "Do you have any idea why that could be problematic? Does it escape you? . . . It's because we have an obligation to—that this project has timelines and if you think the way to manage a project is that at every single decision point you bring it out to a wider group. . . . You guys calling to redesign the memorial seems very problematic at this point, 460 days from its [opening]. . . . My point is that, if you think the decision process should be that once we talk to families that are supposedly representing and then we say actually we're going to disregard that because we haven't asked every single family. That's not a way to move a project."[59]

Ultimately, the families participating in the meeting made it clear that they wanted the memorial and museum to send a letter to all World Trade Center families to ask them what they wanted to have done with the human remains. They believed that the majority view should rule. While they had their own views about the plan for the repository, they stated that their goal was to give all families an informed choice.[60] In the end, while admitting that such a letter would be possible, Daniels avoided making any promises—he was both noncommittal and somewhat dismissive of the idea. In addition to telling them that he would have to think things through, he also said, "Let's just break it down to the very brass tacks, being as honest as possible. The concern in general with opening up big issues, or issues, is what the effect that's going to have, obviously, and we are not in a position—I guess what I'm saying is that we're not going to fully decide it at this meeting, but I would like to continue it. And one of the things I want to express is that the degree of choice, in our perspective, and you guys surely have a differing perspective, it is simply not that we feel that nothing's happened and that the choices are completely unconstrained."[61] Later, he articulated his concern that the mailing would throw off the timeline and potentially lead to demands that were impossible to meet. The family members tried to assuage his concerns by stating that they personally only wanted a very simple, private space away from the crowds of the museum that they could access on their own time, but it was no use.

Seeking Direct Democracy

After waiting in vain for several months for the memorial and museum to contact them, the dissenting families and their advisors took matters into their own hands. In addition to engaging the mass media, they sought permission from Mayor Bloomberg to draft a letter to families explaining what they perceived to be the change in plans regarding the repository and to ask the families whether they approved or disapproved. The dissenting families offered to pay for all expenses associated with the letter, but asked the city to do the mailing so that they did not violate any privacy rules.[62]

When this request was denied, the families worked with attorney Norman Siegel to submit a Freedom of Information Law (FOIL) request to compel the city and the OCME (which maintained the victim database) to release the names and addresses of next of kin of all victims so that they could inform them directly of the plans for the repository.[63] The families also engaged a retired judge to act as a custodian of this information, to assuage concerns about privacy.[64] Both the city and the OCME denied the request for addresses, but they did agree to send a letter to all families informing them that the repository would be located at bedrock between the two footprints within the structure of the museum.[65]

In follow-up discussions, the families demanded that the letter explicitly state that the repository would be accessible through the museum and describe what the experience of visiting the repository would be like. They also wanted the letter to engage family members in the decision-making process and ask them for input and suggestions. Dissenting families were advocating for a "tomb of the unknowns" at street level with a simple inscription along the lines of "Here Rests In Honored Glory The Heroes Known and Unknown of September 11, 2001, and Who Are Known Only to God," which was suggested by Thomas Meehan III.[66] City officials agreed that the letter would clearly explain how the repository would be accessed, but they did not agree to solicit input from families. The letter laying out these facts, which went out to families in mid-October (co-signed by Joe Daniels and Charles Hirsch), stated that the repository's location had been a settled matter since 2006.[67] It is clear that the purpose of the letter was to inform families about what the city and memorial and museum officials wanted to be a closed matter, not to open it to further debate. The letter concluded, "We continue to keep you and the memory of your loved one in our thoughts, and look forward to being in touch with you as we plan to open the Museum."[68]

Dissenting families were far from satisfied with this arrangement, so they appealed the city's FOIL denial. On October 25, 2011, the New York County Supreme Court upheld the denial, arguing that the release of contact information for victims' next of kin would be an unwarranted invasion of privacy, and this decision was upheld on appeal in January 2013.[69] Unable to obtain access to the contact information of all 9/11 families, the dissenting relatives set up a website, http://RespectHumanRemainsAtThe911Memorial.com, on which they solicited family opinions on the matter.

Despite their lack of success in the courtroom, the dissenting families garnered a tremendous amount of attention in the local and national media. Newspapers and TV news reported widely on the situation, and Internet blogs and bulletin boards were full of discussion. The tone and content of these reports and opinions varied widely, with some supporting the families' concerns and demands, and others dismissing them out of hand—suggesting once again that these families needed to move on and stop trying to inhibit progress at the site. The views of a small group of dissenters provided the fodder for the media to report on the controversy and disagreement within the family community. While oppositional op-ed pieces always referred to the families as a "small group," news reports almost never gave any indication that the dissenting families represented only a tiny fraction of the total World Trade Center family population.

While most of the coverage of the repository focused on its placement below ground behind a wall in the museum, even the quotation chosen to adorn the wall of the repository has been called into question. On April 6, 2011, the *New York Times* published an op-ed in which journalist and writer Caroline Alexander argued that the phrase excerpted from Book 9 of Virgil's *Aeneid*—"No Day Shall Erase You From the Memory of Time"—was homoerotic because it referred to the simultaneous battle deaths of two Trojan warriors who were also likely lovers.[70] The pair, Nisus and Euryalus, were killed to avenge the pair's murderous rampage against their enemies. While Alexander understood the appeal of classic texts in situations when our own words seem "hollow and inadequate," she described the Virgil quotation as "disastrous" and "grotesque," and argued that it "dissolve[ed] upon inspection," particularly when read in the context of the *Aeneid*'s storyline.[71]

Alexander's interpretation of the quotation certainly made a splash in the newspaper, and many classicists agree with her assessment.[72] But the borrowing of snippets of musical, artistic, and literary works—often regardless of context—has a long history. Indeed, the Virgil quotation has been used on at least one other monument, the Valiants Memorial in Ottawa, honoring Canadian war heroes, and was widely appropriated in antiquity in funerary contexts as well.[73] Today, the quotation adorns the wall of the repository, although the museum quietly dropped "*Aeneid*" from the tagline to distance the phrase from its origins.[74] Beneath the quotation is a simple plaque that states: "Reposed behind this wall are the remains of many who perished at the World Trade Center site on September 11, 2001. The Repository is maintained by the Office of Chief Medical Examiner of the City of New York."

Supportive Families Respond

In the face of dissenting family activism over the supposed lack of proper consultation with relatives of 9/11 victims, families that had played a central role in the creation of the museum decided they could no longer sit by and risk further delays. In a letter addressed to 9/11 family members, Virginia Bauer, Paula Grant Berry, Mary and Frank Fetchet, Christine Ferer, Monica Iken, Anthoula Katsimatides, Margie Miller, Tom Rogér, and Charles Wolf sought to "set the record straight" on the storage of remains at the World Trade Center site. These relatives, who were all affiliated with the LMDC, the Memorial Foundation, or the National September 11 Memorial and Museum in some way, argued that families had been involved in the decision-making process from the very beginning and that the memorial and museum had merely honored the requests that these families had made throughout. They also argued that the memorial and museum had not acted secretively. It had gone to great lengths to solicit the views of scholars, religious figures, museum ethicists, and others in order to ensure that the remains would be stored in a way that was sensitive, caring, and honest. "We understand and respect that opinions may differ on the issue of how to repose the remains. However, please be assured that the plan being implemented was driven both by the Family Advisory Council and a separate entity of family organizations themselves [the Coalition of 9/11 Families]. And, it was done through a consultative process that considered family members' wishes."[75]

On April 20, 2011, Monica Iken, Thomas Johnson (a member of the WTC Memorial Foundation Board and the LMDC Family Advisory Council), and Charles Wolf published an editorial in the *Wall Street Journal* in which they accused "a small group of 9/11 family members" of distorting the truth—not just about the repository but also about the very demands made by families early throughout the design process.[76] They also rejected calls for a "tomb of the unknown" on two grounds: first, because of the promise made by the medical examiner to retest remains in perpetuity as new methods become available; and, second, because an above-ground tomb would likely become the very magnet of gawking tourists and conspiracy theorist demonstrations that the dissenting families hoped to avoid.

On the first justification, they noted that the additional testing that would be carried out by the medical examiner "makes it necessary to have a working repository instead of a permanent tomb. . . . Unlike at Arlington, where there is no ongoing effort to identify the unknown remains, advances in DNA technology hold out the possibility for identification of the unidentified remains from it."[77] This justification was not entirely accurate since the unidentified remains of the unknown soldier from the Vietnam War had recently been disinterred when they were confirmed to be those of Michael J. Blassie, a pilot shot down in 1972.[78] Further, since the implementation of NAGPRA in 1990, researchers and native groups had worked together in many cases to create storage facilities that were both respectful to the remains and lineal descendants and open to scientists who wished to access the collections.[79] Iken, Johnson, and Wolf concluded by arguing that the National September 11 Memorial and Museum staff had displayed "incredible sensitivity" to the families during the entire planning of the memorial and museum. It had "devoted all of its efforts to make it clear that Ground Zero is not just the site of a deadly attack, it is the final resting place of thousands of our loved ones" and, by virtue of this fact, did not deserve to be accused by a few families of malfeasance.[80]

In what was now a predictable turn of events, dissenting families immediately issued statements expressing their displeasure with the letter to

9/11 families and the op-ed. Maureen and Al Santora, parents of firefighter Christopher Santora, wrote that the statement by the ten families "shows contempt for all of the other 9/11 family members who were never even asked for their opinion" about the remains at the memorial and museum. "We are both outraged and disappointed that those who signed would think that their input was more valuable and important than the other 2,749 families who were not included in this decision."[81] Rather than speak on behalf of the families, the Santoras argued that it was Fetchet's responsibility, as the head of one of the biggest family groups with a large contact list, to inform the other families and to gather their opinions.

Instead, they wanted remains to be stored at the medical examiner's office until such time that it seemed improbable that they would be identified and then placed in a tomb-like structure that was accessible twenty-four hours a day "instead of only during museum hours." Thomas J. Meehan III wrote an open letter to Mary Fetchet, who ran the prominent family group Voices of September 11th, in which he expressed his unhappiness with the notion that those families who worked closely with the memorial and museum were in a position to "set the record straight." It "implies a certain arrogance on the part of the authors, that they are right, and all others are wrong, who oppose them or have different opinions. A poor choice of words for a beginning statement and a document which is meant to inform and convey a sequence of events, and which I assume is also meant to persuade people. One expects more from those who wear the mantle of leadership. For family members who have different opinions, and have not given permission to be represented by those named in your letter, [this] is what I see at the heart of this issue."[82] He stated that he had not been personally informed of, nor had he participated in, the deliberations that took place concerning the remains, despite the fact that he believed that this was his right as a lineal descendent of a victim. "To attempt to justify a process that was flawed to begin with, is without merit, and shame on all of you for believing you had that right to make that decision without consulting the victims' next of kin or gaining their permission and approval."[83]

Ultimately, memorial and museum officials resisted being forced to rethink the placement of the remains. The controversy about human remains would not delay construction after all. Unfortunately for the museum staff, the same could not be said about politics and money, the two constant thorns in the side of the redevelopment effort, or Mother Nature.

Politics, Money, and Floodwater

Shortly after the tenth anniversary of the attacks in 2011, a dispute emerged between the Port Authority and the National September 11 Memorial and Museum over who was responsible for paying for certain aspects of the construction project. The Port Authority argued that the foundation owed it between $150 million and $300 million dollars in construction costs (depending upon the final costs for certain tasks), while the foundation argued that the Port Authority was responsible for picking up the tab. The two sides could not come to an agreement, and construction ground to a halt for more than a year.[84]

For families, this was yet another reminder that the rebuilding of the site was about much more than them. In an interview with Voice of America, Monica Iken said, "It's an embarrassment for the world to see. They come there, and I've been there several times where people come up to me and say, 'Where's the museum, why is it not open?' How do you explain that: 'Oh, because we're fighting over some money?' "[85]

The Port Authority and the memorial and museum finally reached agreement on September 10, 2012, the day before the museum was originally slated to open. In exchange for paying for most cost overruns, the Port Authority gained financial and managerial oversight over the operation of the museum along with a $1 million cut of museum revenues from 2018 to 2048.[86]

Just as construction was gearing back up, Hurricane Sandy devastated the New York region in late October 2012, pouring seven feet of water into the basement of the museum.[87] Dissenting families were incensed that the remains would be stored seventy feet below ground in a site that was prone to flooding, but engineers and contractors at the site insisted that the flooding only occurred because the space was still under construction and had not yet been properly sealed. In any case, the storage of human remains in sealed bags and sealed bins would prevent damage from any flooding that did happen.[88]

Eventually, the water from Sandy was pumped out and the damage caused by the incursion was repaired. The project was put back on track and in spring 2013, officials announced that the museum would open in May 2014. The decision to place remains seventy feet below ground would not be revisited. Families would not be polled to ensure that they were still on board with the foundation's plans, and museum officials expressed confidence that they could, and would, get the job done.[89]

On May 15, 2014, the National 9/11 Memorial and Museum was formally dedicated in a ceremony that included a speech by President Barack Obama and was attended by former President Bill Clinton, former New York Senator Hillary Clinton, former mayors Rudy Giuliani and Michael Bloomberg, recently elected mayor Bill de Blasio, first responders, relatives of victims, and numerous other dignitaries. As he stood before his audience, President Obama proclaimed that the United States is a "nation that stands tall and united and unafraid—because no act of terror can match the strength or the character of our country. Like the great wall and bedrock that embrace us today, nothing can ever break us; nothing can change who we are as Americans."[90] The central message of Obama's speech was that redemption and resilience triumphed over trauma.

Epilogue

On the evening before the National September 11 Memorial and Museum formally opened to the public, donors and corporate supporters were fêted with a cocktail party to recognize their contributions to making the museum possible. For relatives who had labored tirelessly to ensure the creation of a museum that honored the memory of their loved ones, the party was a culmination of more than a decade of advocacy and planning. For relatives who opposed the museum plans and had worked so hard to bring victims' remains above ground, the party was an outrage.[1] The event added to their sense that the people who planned and financed the memorial and museum were more concerned about their own legacies than the feelings of those who lost family and friends on September 11. The dead, it seemed to some, belonged first and foremost to the powerful.

The insult of having victims' remains stored in the basement of the museum, and the perception that those remains were being used to entice tourists to pay for entrance, was too much to bear. For Rosemary Cain, mother of firefighter George Cain, museum officials were "grave robbers" using the remains of World Trade Center victims for "greed and ego," not to uplift the nation or provide solace to those in mourning.[2] For Jim Riches, the presence of remains in the repository would keep him away. "I'll never set foot in that museum," he insisted, "until those remains are out of there and above grade on a plaza in a respectful place where it's open to all the public to go for free."[3]

Thus, in the end, the situation appears to be stuck in a web of competing interests and viewpoints. One can imagine many different outcomes, but there is no simple solution to the storage of the remains of World Trade Center victims that will satisfy everyone. Sites similar to National 9/11 Memorial and Museum—Srebrenica, Auschwitz, Gettysburg, Hiroshima—did not have to weigh the need to memorialize against the need to redevelop and revitalize the surroundings in the stark terms that confronted the stakeholders at the World Trade Center site.[4] While dissenting families offered alternative locations for the storage of remains, the city dismissed them as being impractical.

Ultimately, the dispute about the storage of remains points to the fact that the new scientific identification techniques being developed in the context of the September 11 attacks and other mass fatality events are a double-edged sword. They promise, at least in theory, to eliminate the possibility of identifiable remains being left unidentified. But the same advances in identification techniques mean that remains must be cared for and kept present, both physically and in the minds of surviving relatives and society at large, until they can be identified. In her letter to 9/11 families regarding the transfer of remains to the repository, acting chief medical examiner Barbara A. Sampson, who took over from Charles Hirsch when he retired at the end of 2014, reiterated the OCME's commitment "do whatever it takes for as long as it takes to identify all those who lost their lives."[5] Thus, the remains of the World Trade Center attacks can never fully be laid to rest and become part of "the past."

We may be less prone to forget these remains, and the victims to whom they belong, but sometimes the receding from view of those who have died makes it easier to get on with life. There should not be a "statute of limitations" on mourning or grief, but it is important to at least consider the implications of an identification effort without end. When, if ever, do families and society benefit from saying that the identification phase of an operation is over and that unidentified tissue will remain so indefinitely? In the case of the World Trade Center victims, such a decision would have altered the plan to store remains within the museum because there would have been no requirement that they be permanently accessible by the OCME. It would at least have provided more clarity. "Everything seems jarring and contradictory" at the museum, write historians Charles B. Strozier and Scott Gabriel Knowles. "It will be, but also won't be, a cemetery. . . . And how exactly are visitors expected to behave knowing that the unknown dead from that day remain just behind the wall?"[6]

Unfortunately, there is no satisfying answer to this question. This uncertainty results when grief and mourning intersect with the inherently messy, political process of memorialization. As we have seen, the recovery, identification, and memorialization of the victims of the September 11 World Trade Center attacks certainly could have been improved on many levels. It is unlikely, though, that any alternative would have been satisfactory to everyone.

With time, the political and emotional battles surrounding the remains of the victims of the World Trade Center attacks will recede from view. My goal has been to make sense of these disputes, especially the viewpoints of the relatively small subset of families who disagreed publicly with the actions of city officials and memorial and museum personnel, without passing judgment. Readers will have to decide if these families' concerns were legitimate or not, and if city officials and other powerful figures lived up to their ethical and civic responsibilities. Answers to these questions will depend on each reader's relationship to the events of September 11 and mass violence generally, their education and life experiences, and their personal views on trauma and grief. Perhaps at some future date, the answers to these questions will be more clear-cut, or at least more settled, but that moment has not yet come.

What is clear is that DNA technology has irrevocably altered the way we memorialize the dead, whether their remains have been identified or not. Whereas a tomb of the unknown, a war memorial, or a common burial ground valorizes collective sacrifice to the nation, the repository in the National 9/11 Memorial and Museum is the physical embodiment of the technological dream that unidentified remains may one day be made personal again, and returned to their families. The repository is not a final resting place, even though many of the remains will never be claimed. Rather, it is a storage facility, albeit a dignified one, supporting an ongoing forensic investigation by a government scientific laboratory.

This effort certainly demonstrates the OCME's and the city's ongoing commitment to the victims of the World Trade Center attacks and their families. But nagging questions remain. Are victims' families, their communities, and the nation well-served by the continuation of the identification effort? For some families, continued testing may mean the return of yet another fragment of tissue—perhaps a finger, bone sliver, or a bit of muscle—to go along with the other body parts that have already been returned. The effect of such recoveries on families is impossible to generalize.

But for the families of the 1,113 victims who have not yet received any remains, this unending identification effort may eventually yield the first definitive proof that their loved one perished in the attack, to compliment the administrative determination of death issued in the months after the attack. In March 2015, Matthew David Yarnell, a twenty-six-year-old employee of Fiduciary Trust and a graduate of Carnegie Mellon University, where I teach, became the latest victim to be identified. In an interview with CNN, his mother, Michelle Yarnell, said that the identification initially "opened up all of the old wounds and old pains," but that it ultimately allowed her family to "finally put everything to rest."[7]

Other than the undocumented immigrants never officially reported missing, it is unlikely that any of the victims of the 9/11 World Trade Center attacks suffered an anonymous death. Most victims, including Matthew Yarnell, are named on the parapets around the void and pictured in the museum, and Matthew's mother has said in numerous interviews that they never doubted that he was killed in the attacks. Yet the thought of unidentified remains is unnerving, especially for a society that wants to believe it has the technical capacity to provide some measure of certainty in an uncertain world. Even after Matthew was identified, his mother found solace in the fact that the identification effort would continue in perpetuity. "The ME's office is not going to give up," she said. "I hope for everyone that lost a loved one there, that they'll have that closure someday, and, hopefully, sooner rather than later."[8]

This sentiment seems particularly important in the aftermath of an attack that targeted ordinary people from all walks of life, religions, and national origins, and killed them as representatives of a monolithic United States. It is ironic, then, that the individualization of the victims of the World Trade Center has made it more politically palatable for the U.S. government to engage in a seemingly perpetual war that has created innumerable casualties in Afghanistan, Iraq, Pakistan, Yemen, and elsewhere. One final uncomfortable question we might ask is whether the same kind of unending scientific effort will be taken to identify these victims. The answer highlights the inherently political nature of the deployment of genetic technologies on behalf of the victims of mass atrocity.

Notes

Acknowledgments

Index

Notes

Introduction

1. Office of the Chief Medical Examiner, "World Trade Center Operational Statistics," 2015, http://www.nyc.gov/html/ocme/downloads/pdf/public_affairs_ocme_pr_WTC_Operational_Statistics.pdf.

2. Simon Robins, *Families of the Missing: A Test for Contemporary Approaches to Transitional Justice* (New York: Routledge, 2013); ICRC, "Report from the International Conference on The Missing and Their Families," February 19–21, 2003, http://www.icrc.org/eng/assets/ files/other/icrc_002_0857.pdf.

3. Victor Toom, "Whose Body Is It? Technolegal Materialization of Victims' Bodies and Remains after the World Trade Center Terrorist Attacks," *Science, Technology, and Human Values*, forthcoming.

4. As Claire Moon notes, "forensic knowledge does not finalise, but interacts with social, political and historical interpretations of past violence in ways that are both conflicted and unpredictable." Moon, "Interpreters of the Dead: Forensic Knowledge, Human Remains and the Politics of the Past," *Social and Legal Studies* 22, no. 2 (2012): 149.

5. I do not consider Pearl Harbor here because Hawaii was a U.S. territory, and not a state, during World War II. I discuss the USS *Arizona* Memorial below.

6. Avner Ben-Amos, *Funerals, Politics, and Memory in Modern France 1789–1996* (Oxford: Oxford University Press, 2000); Thomas Laqueur, "Spaces of the Dead," *Ideas: The National Humanities Center* 8, no. 2 (2001): 3–16; István Rév, *Retroactive Justice: Prehistory of Post-Communism* (Stanford, CA: Stanford University Press, 2005); Katherine Verdery, *The Political Lives of Dead Bodies: Reburial and Postsocialist Change* (New York: Columbia University Press, 1999).

7. Mamphela Ramphele, "Political Widowhood in South Africa: The Embodiment of Ambiguity," *Daedalus* 125, no. 1 (1996): 99–117.

8. Rita Arditti, *Searching for Life: The Grandmothers of the Plaza de Mayo and the Disappeared Children of Argentina* (Berkeley: University of California Press, 1999); Jay D. Aronson, "The Strengths and Limitations of South Africa's Search for Apartheid-Era Missing Persons," *International Journal of Transitional Justice* 5, no. 2 (2011): 262–281; Anna Eyre and Pam Dix, *Collective Conviction: The Story of Disaster Action* (Liverpool: Liverpool University Press, 2015); Jo Fisher, *Mothers of the Disappeared* (Boston: South End Press, 1989); Robins, *Families of the Missing*; Victoria Sanford, *Buried Secrets: Truth and Human Rights in Guatemala* (New York: Palgrave Macmillan, 2003); Sarah Wagner, *To Know Where He Lies: DNA Technology and the Search for Srebrenica's Missing* (Berkeley: University of California Press, 2008).

9. Fisher, *Mothers of the Disappeared.*

10. Wagner, *To Know Where He Lies.*

11. Aronson. "Strengths and Limitations"; Shari Eppel, "'Healing the Dead': Exhumation and Reburial as Truth-Telling and Peace-Building Activities in Rural Zimbabwe," in *Telling the Truths: Truth Telling and Peace Building in Post-Conflict Societies*, ed. Tristan Anne Borer (Notre Dame, IN: University of Notre Dame Press, 2006), 259–288.

12. Wagner, *To Know Where He Lies;* Sarah Wagner, "The Making and Unmaking of an Unknown Soldier," *Social Studies of Science* 43, no. 5 (2013): 631–656.

13. Adam Rosenblatt, *Digging for the Disappeared: Forensic Science after Atrocity* (Stanford, CA: Stanford University Press, 2015).

14. Kenneth E. Foote, *Shadowed Ground: America's Landscapes of Violence and Tragedy,* rev. ed. (Austin: University of Texas Press, 2003).

15. A memorial to the victims of the Pearl Harbor attacks was dedicated in 1962, and a visitors' center opened in 1980. National Park Service, "World War II Valor in the Pacific Frequently Asked Questions," n.d., http://www.nps.gov/valr/faqs.htm.

16. Edward T. Linenthal, *The Unfinished Bombing: Oklahoma City in American Memory* (New York: Oxford University Press, 2001).

17. George L. Mosse, *Fallen Soldiers: Reshaping the Memory of the World Wars* (Oxford: Oxford University Press, 1990); Jay M. Winter, *Sites of Memory, Sites of Mourning: The Great War in European Cultural History* (Cambridge: Cambridge University Press, 1995).

18. The Civil War battle of Gettysburg was a prominent exception to this rule.

19. Thomas W. Laqueur, "Names, Bodies, and the Anxiety of Erasure," in *The Social and Political Body*, ed. Theodore R. Schatzki and Wolfgang Natter (New York: Guilford Press, 1996), 123–161.

20. Ibid., 132.

21. Wagner, "Making and Unmaking."

22. Lacquer, "Names, Bodies." More recently, anthropologist Marilyn Strathern argues that in Western culture the body and its genetic blueprint have become the way that we represent and think of the individual. Marilyn Strathern, *Kinship, Law and the Unexpected: Relatives Are Always a Surprise* (Cambridge: Cambridge University Press, 2005).

23. Benedict Anderson, *Imagined Communities: Reflections on the Origin and Spread of Nationalism,* rev. ed. (New York: Verso, 1991); Geoff Dyer, *The Missing of the Somme* (New York: Vintage, 2011); Paul Fussell, *The Great War and Modern Memory* (Oxford: Oxford University Press, 1975); John R. Gillis, ed. *Commemorations: The Politics of National Identity* (Princeton, NJ: Princeton University Press, 1994); Mosse, *Fallen Soldiers*; Winter, *Sites of Memory.*

24. David Simpson, *9/11: The Culture of Commemoration* (Chicago: University of Chicago Press, 2006).

25. "Among the Missing," *New York Times*, October 14, 2001.

26. Robert C. Shaler, *Who They Were: Inside the World Trade Center DNA Story; The Unprecedented Effort to Identify the Missing* (New York: Free Press, 2005); Eve Conant, "Terror: The Remains of 9/11 Hijackers," *Newsweek*, January 2, 2009, http://www.newsweek.com/terror-remains-911-hijackers-78327.

27. Riva Kastoryano, *Que faire des corps des djihadistes? Territoire et identité* (Paris: Fayard, 2015).

28. Chris Lawrence, "'No Land Alternative' Prompts bin Laden Sea Burial," CNN, May 2, 2011, http://www.cnn.com/2011/WORLD/asiapcf/05/02/bin.laden.burial.at.sea/. By contrast, in Israel, bodies of Palestinian suicide bombers who blew themselves up during the Second Intifada were, until recently, buried in special cemeteries in unmarked graves, but they are now being exhumed and handed over to families. Reuters, "IDF Confirms Israel Returning Remains of Palestinian Terrorists," *Jerusalem Post*, January 20, 2014, http://www.jpost.com/Middle-East/IDF-confirms-Israel-returning-remains-of-Palestinian-terrorists-338719; Associated Press, "Israel Returns Palestinian Bombers' Bodies, Decade After Attacks," *Haaretz*, February 2, 2014, http://www.haaretz.com/israel-news/1.572061.

29. Lawrence, "No Land Alternative."

30. Kastoryano, *Que faire des corps des djihadistes?*

31. Robin Wagner-Pacifici and Barry Schwartz, "The Vietnam Veterans Memorial: Commemorating a Difficult Past," *American Journal of Sociology* 97, no. 2 (1991): 376–420.

32. Linenthal, *Unfinished Bombing*, 228.

33. Wagner-Pacifici and Schwartz, "The Vietnam Veterans Memorial."

34. Ronald N. Giere, "Controversies Involving Science and Technology: A Theoretical Perspective," in *Scientific Controversies: Case Studies in the Resolution and Closure of Disputes in Science and Technology*, ed. H. Tristram Englehardt Jr. and Arthur L. Caplan (Cambridge: Cambridge University Press, 1987), 125–150; Bruno

Latour, *Reassembling the Social: An Introduction to Actor-Network Theory,* Clarendon Lectures in Management Studies (Oxford: Oxford University Press, 2005).

35. Marita Sturken, *Tourists of History: Memory, Kitsch, and Consumerism from Oklahoma City to Ground Zero* (Durham, NC: Duke University Press, 2007); Mateo Taussig-Rubbo, "Sacred Property: Searching for Value in the 9/11 Rubble," Religion & Culture Web Forum, Martin Marty Center for the Advanced Study of Religion, University of Chicago (2009), https://divinity.uchicago.edu/sites/default/files/imce/pdfs/webforum/062009/sacred%20property%20final.pdf.

36. Elizabeth Greenspan, *Battle for Ground Zero: Inside the Political Struggle to Rebuild the World Trade Center* (New York: Palgrave Macmillan, 2013).

37. Durkheim (1858–1917) is considered a founder of the modern discipline of sociology, He is particularly well known for his view that society and culture emerge through the interactions of the individuals that compose them, and do not supersede these interactions.

38. Paul Williams, *Memorial Museums: The Global Rush to Commemorate Atrocities* (New York: Berg, 2007).

39. Diane Cardwell, "A Nation Challenged: Ground Zero; First Viewing Platform Opens to the Public," *New York Times*, December 30, 2001, B8.

40. Ibid.

41. Linenthal, *Unfinished Bombing*, 234.

42. For a political analysis of the reading of the Gettysburg address on this occasion, see Simon Stow, "Pericles at Gettysburg and Ground Zero: Tragedy, Patriotism, and Public Mourning," *American Political Science Review* 101, no. 2 (2007): 195–208.

43. Southern Poverty Law Center, "Anti-Muslim Incidents Since Sept. 11, 2001," 2011, https://www.splcenter.org/news/2011/03/29/anti-muslim-incidents-sept-11-2001.

44. Simpson, *9/11.*

45. Sturken, *Tourists of History*, 202.

46. Ibid., 203.

47. Robert Kolker, "The Grief Police," *New York Magazine*, November 28, 2005, http://nymag.com/nymetro/news/sept11/features/15140; Jennifer Senior, "The Memorial Warriors," *New York Magazine,* September 15, 2002, http://nymag.com/nymetro/news/sept11/features/n_7691/.

48. Deborah Sontag, "Broken Ground the Hole in the City's Heart," *New York Times*, September 11, 2006, special section, 1–10.

49. Linenthal, *Unfinished Bombing.*

50. Ibid., 43.

51. Ibid., 96.

52. Ibid., 196.

53. Wagner, *To Know Where He Lies*, 11.

54. Wagner, *To Know Where He Lies.*

55. Erika Doss, *Memorial Mania: Public Feeling in America* (Chicago: University of Chicago Press, 2010); Foote, *Shadowed Ground*; Linenthal, *Unfinished Bombing*; Sturken, *Tourists of History*; Winter, *Sites of Memory.*

56. J. Shawn Landres and Oren Baruch Stier, "Introduction," in *Religion, Violence, Memory, and Place,* ed. Oren Baruch Stier and J. Shawn Landres (Bloomington: Indiana University Press, 2006), 1–12.

57. Doss, *Memorial Mania*; Sturken, *Tourists of History.*

58. The USS *Arizona* Memorial in Pearl Harbor took more than twenty years build, with a visitors' center being constructed only in 1980. Geoffrey M. White, "National Subjects: September 11 and Pearl Harbor," *American Ethnologist* 31, no. 3 (2004): 293–310.

59. Doss, *Memorial Mania*; Linenthal, *Unfinished Bombing.*

1: A Tuesday Morning in September

1. National Institute of Standards and Technology, *Final Report on the Collapse of the World Trade Center Towers* (Gaithersburg, MD: NIST, 2005). http://www.nist.gov/customcf/get_pdf.cfm?pub_id=909017.

2. Ibid.

3. New York City Fire Department, "McKinsey Report: Increasing FDNY's Preparedness," NYC.gov, 2002, http://www.nyc.gov/html/fdny/html/mck_report/toc.html.

4. NIST, *Final Report.*

5. NYFD, "McKinsey Report."

6. Ibid.

7. NIST, *Final Report.*

8. NYFD, "McKinsey Report."

9. Ibid.

10. National Commission on Terrorist Attacks Upon the United States, *The 9/11 Report* (New York: St. Martins, 2004).

11. NYFD, "McKinsey Report."

12. NIST, *Final Report.*

13. James Luongo (NYPD inspector), interview with Jamie York and Elinoar Astrinsky, June 19, 2002, 9/11 World Trade Center Collection, New York State Museum and Archive.

14. Damon DiMarco, *Tower Stories: An Oral History of 9/11* (Santa Monica, CA: Santa Monica Press, 2007), 35.

15. Dennis Cauchon and Martha Moore, "Desperation Forced a Horrific Decision," *USA Today*, September 2, 2002; Kevin Flynn and Jim Dwyer, "Falling Bodies, a 9/11 Image Etched in Pain," *New York Times*, September 10, 2004, A1.

16. Tom Junod, "The Falling Man," *Esquire,* September 2003, http://classic.esquire.com/the-falling-man/.

17. Amy Mundorff (forensic anthropologist injured in collapse of South Tower), interview with author, January 19, 2012, Knoxville, TN.

18. Ibid.

19. Timothy Burke (firefighter), "World Trade Center Task Force Interview," January 22, 2002, http://graphics8.nytimes.com/packages/pdf/nyregion/20050812_WTC_GRAPHIC/9110488.PDF.

20. See, for example, Sam Melisi (member of Rescue 3, FDNY Building Collapse Unit, the Bronx), interview, 2002, *America Rebuilds: A Year at Ground Zero,* PBS, http://www.pbs.org/americarebuilds/profiles/profiles_melisi_t.html.

21. William Langewiesche, *American Ground: Unbuilding the World Trade Center* (New York: North Point Press, 2003), 8.

22. Richard Garlock (engineer who advised on structural issues at WTC), interview, 2002, *America Rebuilds: A Year at Ground Zero,* PBS, http://www.pbs.org/americarebuilds/profiles/profiles_garlock_t.html.

23. NYFD, "McKinsey Report."

24. Gary Suson, "Band of Dads Virtual Exhibit," Ground Zero Museum Workshop, accessed March 9, 2016, http://www.groundzeromuseumworkshop.com/band/dad.htm.

25. Melisi, interview, *America Rebuilds.*

26. City Lore, "Missing: A Streetscape of a City in Mourning," accessed March 9, 2016, http://citylore.org/wp-content/uploads/2011/10/911_exhibit/911.html; Martha Cooper, "Remembering 9/11" (2011), http://marthacooper.viewbook.com/remembering_911/album/wtc__missing_posters?p=1.

27. Robert N. Munson, "September 11 Digital Archive: Personal Memories" (2001). http://911digitalarchive.org/items/show/422.

28. Robert C. Shaler, *Who They Were: Inside the World Trade Center DNA Story: The Unprecedented Effort to Identify the Missing* (New York: Free Press, 2005).

29. Mike Hennessey, "World Trade Center DNA Identifications: The Administrative Review Process," paper presented at the Proceedings of the Thirteenth International Symposium on Human Identification (2002), https://www.promega.com/~/media/files/resources/conference%20proceedings/ishi%2013/oral%20presentations/hennesseyrev1.pdf?la=en; Shiya Ribowsky and Tom Shachtman, *Dead Center: Behind the Scenes at the World's Largest Medical Examiner* (New York: Harper Collins, 2006); Shaler, *Who They Were.*

30. Shaler, *Who They Were.*

31. Ribowsky and Shachtman, *Dead Center,* 225.

32. Dan Barry, "A Few Moments of Hope in a Mountain of Rubble," *New York Times,* September 13, 2001, A1.

33. Ribowsky and Shachtman, *Dead Center.*

34. Dan Barry, "With No Miracle in the Rubble, Hope Grimly Shifts to Acceptance," *New York Times*, September 17, 2001, A12.

35. President's DNA Initiative, *Lessons Learned From 9/11: DNA Identification in Mass Fatality Incidents* (Washington, DC: National Institute of Justice, 2006).

36. Mundorff, interview.

37. Langewiesche, *American Ground*, 96.

38. DiMarco, *Tower Stories*.

39. Garlock, interview, *America Rebuilds*.

40. William Keegan Jr., *Closure: The Untold Story of the Ground Zero Recovery Mission*, with contributions by Bart Davis (New York: Touchstone, 2006), 82.

41. Keegan, *Closure*.

42. Ibid.

43. Langewiesche, *American Ground*.

44. Keegan, *Closure*.

45. Langewiesche, *American Ground*.

46. Ibid.

47. Ibid., 72.

48. Keegan, *Closure*, 83–84.

49. Greg Gittrich, "Mournful Task at an End. Ground Zero Search Over," *New York Daily News*, May 29, 2002, 2.

50. Langewiesche, *American Ground*.

51. Charlie Vitchers (Bovis Lend Lease superintendent), interview, 2002, *American Rebuilds: A Year at Ground Zero*, PBS, http://www.pbs.org/americarebuilds/profiles/profiles_vitchers_t.html.

52. DiMarco, *Tower Stories*, 256.

53. Langewiesche, *American Ground*, 69.

54. Ibid.

55. Michelle McPhee, "Rescuers Honor Each Grim Find in Rubble," *New York Daily News*, November 3, 2001, 2.

56. Mundorff, interview.

57. Amy Mundorff, "Anthropologist-Directed Triage: Three Distinct Mass Fatality Events Involving Fragmentation of Human Remains," in *Recovery, Analysis and Identification of Commingled Human Remains*, ed. Bradley J. Adams and John E. Byrd (Totowa, NJ: Humana Press, 2008), 123–144.

58. Mundorff, interview.

59. Amy Mundorff, "Human Identification Following the World Trade Center Disaster: Assessing Management Practices for Highly Fragmented and Commingled Human Remains," (PhD diss., Simon Frasier University, 2009), http://summit.sfu.ca/item/9782.

60. Richard A. Gould, *Disaster Archaeology* (Salt Lake City: University of Utah Press, 2007).

61. Mundorff, "Human Identification," 105–106.

62. Mundorff, "Human Identification."

63. Gould, *Disaster Archaeology.*

64. Ibid.

65. Bob Port and Joe Calderone, "Satellite Map of Tragedy. Remains, Belongings, Plotted at WTC Site," *New York Daily News*, March 26, 2002, 8. Original version on file with author.

66. Gould, *Disaster Archaeology*, 45.

67. Amy Waldman, "With Solemn Detail, Dust of Ground Zero Is Put in Urns," *New York Times*, October 15, 2001, B11.

68. Ibid.

69. Dan Barry and Kevin Flynn, "Firefighters in Angry Scuffle with Police at Trade Center," *New York Times*, November 3, 2001, A1.

70. Charlie LeDuff, "As Dig Goes on, Emotions Are Buried Deep," *New York Times*, November 18, 2001, B1.

71. Quoted in Jennifer Steinhauer, "A Nation Challenged: The Mayor; Blunt Words to Describe Grim Reality at the Site," *New York Times*, November 3, 2001, B10.

72. Ibid.

73. Langewiesche, *American Ground.*

74. Barry and Flynn, "Firefighters in Angry Scuffle."

75. Jennifer Steinhauer, "Mayor Criticizes Firefighters Over Stand on Staffing At Trade Center Site," *New York Times*, November 9, 2001, D5.

76. Robert D. McFadden, "Second Union Leader is Charged with Trespassing in Demonstration at Ground Zero," *New York Times*, November 5, 2001.

77. Rita Arditti, *Searching for Life: The Grandmothers of the Plaza de Mayo and the Disappeared Children of Argentina* (Berkeley: University of California Press, 1999); Sarah Wagner, *To Know Where He Lies: DNA Technology and the Search for Srebrenica's Missing* (Berkeley: University of California Press, 2008).

78. Dan Barry, "As Sept. 11 Widows Unite, Grief Finds Political Voice," *New York Times*, November 25, 2001, A1.

79. Marian Fontana, *A Widow's Walk: A Memoir of 9/11* (New York: Simon and Schuster, 2005).

80. Barry, "As Sept. 11 Widows Unite."

81. Give Your Voice, "About Give Your Voice for WTC Victims & Families," 2001, http://www.giveyourvoice.bizland.com/aboutus.html.

82. Give Your Voice, "The Forgotten Civilians," November 17, 2001, http://www.giveyourvoice.bizland.com/november17.html.

83. Ibid.

84. Give Your Voice, "Give Your Voice Latest Information," December 10, 2001, http://www.giveyourvoice.bizland.com/december10.html.

85. Fontana, *Widow's Walk*; Give Your Voice, "November 12, 2001," November 12, 2001, http://www.giveyourvoice.bizland.com/november12.html.

86. Langewiesche, *American Ground,* 168.

87. Monica Iken, interview, 2002, *America Rebuilds: A Year at Ground Zero,* PBS, http://www.pbs.org/americarebuilds/profiles/profiles_iken_t.html.

88. Barry, "As Sept. 11 Widows Unite;" Give Your Voice, "November 12, 2001."

89. Barry, "As Sept. 11 Widows Unite."

90. Richard Pyle, "World Trade Center Families Are Being Warned That Some Victims Were 'Vaporized,'" Associated Press, December 4, 2001.

91. Jennifer Steinhauer, "Ex-Firefighter's Quiet Plea Ends Conflict over Staffing," *New York Times,* November 17, 2001, B9.

92. Gittrich, "Mournful Task at an End."

93. Peter Rinaldi (Port Authority engineer), interview, 2002, *America Rebuilds: A Year at Ground Zero,* PBS, http://www.pbs.org/americarebuilds/profiles/profiles_rinaldi_t.html.

94. Ibid.

2: Fresh Kills

1. Dennis Diggins, affidavit, October 5, 2006. World Trade Center Families for a Proper Burial vs. City of New York et al., No. 08-3705-CV (U.S. Ct. App. 2d Cir., 2008), Joint Appendix, A-74 to A-80. On file with author.

2. William Langewiesche, *American Ground: Unbuilding the World Trade Center,* repr, (New York: North Point Press, 2003).

3. James Luongo, affidavit, October 5, 2006. World Trade Center Families for a Proper Burial vs. City of New York et al., No. 08-3705-CV (U.S. Ct. App. 2d Cir., 2008), Joint Appendix, A-59 to A-68. On file with author.

4. The Disaster Mortuary Operational Response Team is a network of experts in mortuary sciences and identification techniques that is activated to respond to mass disasters by the federal government.

5. New York State Museum, "Recovery: The World Trade Center Recovery Operation at Fresh Kills (a Traveling Exhibition from the New York State Museum)" n.d., http://www.nysm.nysed.gov/exhibits/longterm/documents/recovery.pdf.

6. Dan Barry, "Sifting the Last Tons of Sept. 11 Debris," *New York Times,* May 14, 2002, A1; New York State Museum, "Recovery."

7. James Luongo, interview with Jamie York and Elinoar Astrinsky, June 19, 2002, New York State Museum and Archive, 9/11 World Trade Center Collection, Tape 1, side B.

8. Ibid., Tape 1 side A.

9. "WTC Living History Project Group Response—Part I," WTC Living History Project, accessed July 21, 2011, http://www.wtclivinghistory.org/groundzerocorrection1.htm, 8–9.

10. "WTC Living History Project Group Response—Part III," WTC Living History Project, accessed July 21, 2011, http://www.wtclivinghistory.org/groundzerocorrection3.htm.

11. Give Your Voice, "Give Your Voice Latest Information," December 10, 2001, http://www.giveyourvoice.bizland.com/december10.html.

12. Richard Pyle, "At a New York City Landfill for Attack Debris, 'God's Work' Is Under Way," Associated Press, January 15, 2002.

13. Diane Horning, affidavit, March 20, 2007. World Trade Center Families for a Proper Burial vs. City of New York et al., No. 08-3705-CV (U.S. Ct. App. 2d Cir., 2008), Joint Appendix, A-107 to A-113. On file with author.

14. Ibid., 2.

15. Ibid., 4.

16. Amended complaint, June 26, 2006, 6. World Trade Center Families for a Proper Burial vs. City of New York et al., No. 08-3705-CV (U.S. Ct. App. 2d Cir., 2008), Joint Appendix, A-31 to A-52. On file with author.

17. Mateo Taussig-Rubbo, "Sacred Property: Searching for Value in the 9/11 Rubble," Religion & Culture Web Forum, Martin Marty Center for the Advanced Study of Religion, University of Chicago (2009), 24, https://divinity.uchicago.edu/sites/default/files/imce/pdfs/webforum/062009/sacred%20property%20final.pdf.

18. Ibid., 5; Horning, affidavit.

19. Michael Mucci, affidavit, October 2, 2006. World Trade Center Families for a Proper Burial vs. City of New York et al., No. 08-3705-CV (U.S. Ct. App. 2d Cir., 2008), Joint Appendix, A-69 to A-73. On file with author.

20. Amended complaint, 8; Theodore Feasor, affidavit, March 19, 2007, 2. World Trade Center Families for a Proper Burial vs. City of New York et al., No. 08-3705-CV (U.S. Ct. App. 2d Cir., 2008), Joint Appendix, A-165 to A-169. On file with author.

21. Mucci, affidavit.

22. Amy Mundorff, "Human Identification Following the World Trade Center Disaster: Assessing Management Practices for Highly Fragmented and Commingled Human Remains," (PhD diss., Simon Frasier University, 2009), http://summit.sfu.ca/item/9782.

23. Amended complaint, 9.

24. James Taylor, affidavit, March 20, 2007, 2. World Trade Center Families for a Proper Burial vs. City of New York et al., No. 08-3705-CV (U.S. Ct. App. 2d Cir., 2008), Joint Appendix, A-152 to A-156. On file with author. Plaintiffs calculated this number as follows: "Of the approximately 1.651 million tons of debris recovered from the World Trade Center, only approximately 1.05 million tons of such debris was put

through the Taylor and Yannuzzi recycling process. Of the remaining approximately 601,000 tons, approximately 187,000 tons of debris consisted of steel, leaving approximately 414,000 tons of such debris unsifted." (Amended complaint, 9.)

25. Taylor, affidavit, 4.

26. Horning, affidavit, 4.

27. Christy Ferer, letter to World Trade Center Families for a Proper Burial, July 2002. Quoted in World Trade Center Families for a Proper Burial vs. City of New York et al., amended complaint, June 26, 2006, 11–12. On file with author.

28. Susan Edelman, "Anguish Over Loved Ones' Ashes; 9/11 Kin Rip 'Heartless' Bloomberg," *New York Post*, June 22, 2003, 4; Diane Horning (mother of WTC victim and founder of WTCFPB), interview with New York State Museum, February 6, 2008, New York State Museum and Archive.

29. Edelman, "Anguish Over Loved Ones' Ashes."

30. Horning, interview, part 1.

31. Christy Ferer, "Lives Lost and the Renewal of Downtown," *New York Times*, May 18, 2002, A15.

32. Monica Iken, interview with author, July 26, 2011, Manhattan.

33. Martha T. Moore, "WTC Families Want Remains Out of Landfill," *USA Today*, October 6, 2004, 13A.

34. Horning, affidavit, 6.

35. Horning, affidavit.

36. Luongo, affidavit, 3. Department of Sanitation officials Michael Mucci and Dennis Diggins made the same declaration in their affidavits.

37. Luongo, affidavit, 3.

38. Diggins, affidavit; Scott Orr, affidavit, September 22, 2006. World Trade Center Families for a Proper Burial vs. City of New York et al., No. 08-3705-CV (U.S. Ct. App. 2d Cir., 2008), Joint Appendix, A-81 to A-84. On file with author.

39. Diane Horning, e-mail to author, August 1, 2014. On file with author.

40. Amended complaint, 12.

41. Ibid., 13.

42. Patricia Yaeger, "Rubble as Archive, or 9/11 as Dust, Debris, and Bodily Vanishing," in *Trauma At Home: After 9/11*, ed. Judith Greenberg (Lincoln: University of Nebraska Press, 2003), 187–194.

43. Anthony DePalma, "Landfill Park . . . Final Resting Place? Plans for Fresh Kills Trouble 9/11 Families Who Sense Loved Ones in the Dust," *New York Times*, June 14, 2004, B1.

44. This episode is particularly interesting when one contrasts the constructed sacredness of the World Trade Center ash, which was placed in urns and given to families by city officials in October 2001, with the denial of the sacredness of the fines at Fresh Kills.

45. Diane Stewart, e-mail to Diane Horning, October 20, 2002. World Trade Center Families for a Proper Burial vs. City of New York et al., No. 08-3705-CV (U.S. Ct. App. 2d Cir., 2008), Joint Appendix, 150. On file with author.

46. David Saltonstall, "Park Is Mike's Fresh Idea for S.I. Dump," *New York Daily News*, September 30, 2003, 8.

47. Amended complaint.

48. Ibid., 14.

49. Joe Mahoney, "Families: Remove Remains from Fresh Kills," *New York Daily News*, June 16, 2005, 24.

50. Ibid.

51. DePalma, "Landfill Park."

52. Amended complaint.

53. Norman Siegel, letter to Mayor Michael R. Bloomberg, March 11, 2005, 2. World Trade Center Families for a Proper Burial vs. City of New York et al., No. 08-3705-CV (U.S. Ct. App. 2d Cir., 2008), Joint Appendix, A-53 to A-56. On file with author.

54. Siegel, letter to Bloomberg.

55. David Epstein, "WTC Victims' Kin Make Pitch," *New York Daily News*, September 2, 2004, 40.

56. "Respecting 9/11's Dead," *New York Post*, October 19, 2004, 30.

57. Andrea Peyser, "Public Weary of Kin Who Can't Let Go," *New York Post*, October 18, 2005, 24.

58. Air Transportation Safety and System Stabilization Act of 2001, Pub. L. 107-42, 115 Stat. 230 (2001).

59. Amended complaint, 3.

60. Ibid., 4.

61. Ibid.

62. Mireya Navarro, "Empathetic Judge in 9/11 Suits Seen by Some as Interfering," *New York Times*, May 2, 2010, A17.

63. Julia Preston, "Wife of Trade Center Victim Verbally Confronts Judge," *New York Times*, January 14, 2006, B5.

64. Eric Beck, affidavit, March 19, 2007, 3. World Trade Center Families for a Proper Burial vs. City of New York et al., No. 08-3705-CV (U.S. Ct. App. 2d Cir., 2008), Joint Appendix, A-160 to A-164. On file with author.

65. Ibid.

66. John Barrett, affidavit, March 8, 2007, 3. World Trade Center Families for a Proper Burial vs. City of New York et al., No. 08-3705-CV (U.S. Ct. App. 2d Cir., 2008), Joint Appendix, A-173 to A-176. On file with author.

67. Ibid., 4.

68. Feasor, affidavit.

69. Ibid., 5.

70. Beck, affidavit, 4.

71. Charles S. Hirsch, letter to Diane Horning re: presence of human remains in material at Fresh Kills, January 9, 2003. World Trade Center Families for a Proper Burial vs. City of New York et al., No. 08-3705-CV (U.S. Ct. App. 2d Cir., 2008), Joint Appendix, A-180. On file with author.

72. Dennis Diggins, supplemental affidavit, June 20, 2007, 2. World Trade Center Families for a Proper Burial vs. City of New York et al., No. 08-3705-CV (U.S. Ct. App. 2d Cir., 2008), Joint Appendix, A-181 to A-186. On file with author.

73. Ibid., 5–6.

74. Fresh Kills operations meeting minutes held in June 2002. World Trade Center Families for a Proper Burial vs. City of New York et al., No. 08-3705-CV (U.S. Ct. App. 2d Cir., 2008), Joint Appendix, A-196 to A-209. On file with author.

75. Plaintiff's statement of undisputed material facts in opposition to defendants' motion for summary judgment, February 20, 2008. World Trade Center Families for a Proper Burial vs. City of New York et al., No. 08-3705-CV (U.S. Ct. App. 2d Cir., 2008), Joint Appendix, A-210 to A-242. On file with author.

76. Horning, e-mail to author.

77. WTC Families for Proper Burial, "Petition to: Governor Pataki, Mayor Bloomberg & Members of the City Council" (n.d.). World Trade Center Families for a Proper Burial vs. City of New York et al., No. 08-3705-CV (U.S. Ct. App. 2d Cir., 2008), Joint Appendix, A-115 to A-116. On file with author.

78. David Seifman, "Mike: No Fresh Fresh Kills Hunt," *New York Post*, November 4, 2006, 2.

79. Transcript of oral argument proceedings, February 28, 2008, 2. World Trade Center Families for a Proper Burial vs. City of New York et al., No. 08-3705-CV (U.S. Ct. App. 2d Cir., 2008), Joint Appendix, A-243 to A-273. On file with author.

80. Ibid.

81. Ibid., 8.

82. Ibid.

83. Ibid., 10.

84. Ibid., 12.

85. Ibid.

86. Ibid., 14.

87. Ibid., 17.

88. Ibid.

89. Ibid., 18.

90. Ibid., 23.

91. Ibid., 30.

92. World Trade Center Families for a Proper Burial et al. v. City of New York et al., 567 F. Su2d 529, 532–533 (2008).

93. Ibid., 539.

94. Ibid., 536.

95. Ibid., 542.

96. Summary order, U.S. Ct. App., 359 Fed. Appx. 177 (2d Cir. 2009).

97. Ibid., 181.

98. U.S. Supreme Court Docket File 09-1467.

3: Identifying the Dead

1. Shiya Ribowsky, phone interview with author, February 12, 2014.

2. Shiya Ribowsky and Tom Shachtman, *Dead Center: Behind the Scenes at the World's Largest Medical Examiner* (New York: Harper Collins, 2006), 190.

3. Ibid.

4. Ibid.

5. Ibid.; Amy Mundorff, interview with author, January 19, 2012, Knoxville, TN.

6. Ribowsky and Shachtman, *Dead Center*, 158–159.

7. Mundorff, interview.

8. Amy Mundorff, "Human Identification Following the World Trade Center Disaster: Assessing Management Practices for Highly Fragmented and Commingled Human Remains." PhD diss., Simon Frasier University, 2009. http://summit.sfu.ca/item/9782.

9. Mundorff, interview.

10. Mundorff, "Human Identification."

11. Gaille MacKinnon and Amy Z. Mundorff, "World Trade Center - September 11th, 2001," in *Forensic Human Identification: An Introduction*, ed. Timothy Thompson and Susan Black (Boca Raton, FL: CRC Press, 2006), 485–499.

12. Amy Mundorff, "Human Identification."

13. Uriel Heilman, "Dept. of Remembrance: Watching over the 9/11 Dead with *Shmira*," Jewish Telegraphic Agency, September 1, 2011, http://www.jta.org/2011/09/01/life-religion/dept-of-remembrance-watching-over-the-911-dead-with-shmira.

14. Reverend Charles Flood (Episcopal priest who held services for victim's families), phone interview with author, February 6, 2014.

15. Lynn Castrianno, "New York City Trip (Reflections on Visit to World Trade Center Site and OCME Facility), with Commentary by Close Friends," April 2002. Personal collection of Lynn Castrianno. On file with author.

16. JoAnne Wasserman, "Revamped Park Serves 9/11 Families," *New York Daily News*, December 24, 2002, 10.

17. David W. Dunlap, "Renovating a 'Sacred Space,' Where the 9/11 Remains Wait," *New York Times*, August 29, 2006, B3.

18. Ribowsky and Shachtman, *Dead Center*, 5. According to Robert Shaler, OCME staff member John Snyder used a similar metaphor in a conference call in

2002. Shaler, *Who They Were: Inside the World Trade Center DNA Story; The Unprecedented Effort to Identify the Missing* (New York: Free Press, 2005).

19. Shaler, *Who They Were.*

20. Office of Chief Medical Examiner, "World Trade Center Operational Statistics," July 1, 2013, http://www.where-to-turn.org/phpBB2/download.php?id=1859.

21. Margaret Talbot, "The Lives They Lived: 3,225 (At Last Count)," *New York Times Magazine*, December 30, 2001, 16.

22. Shaler, *Who They Were.*

23. Ribowsky and Shachtman, *Dead Center.*

24. Jay D. Aronson, *Genetic Witness: Science, Law, and Controversy in the Making of DNA Profiling* (New Brunswick, NJ: Rutgers University Press, 2007).

25. Robert Lee Hotz, "Probing the DNA of Death," *Los Angeles Times*, October 9, 2002, http://articles.latimes.com/2002/oct/09/science/sci-remains9.

26. Shaler, *Who They Were.*

27. Ibid.

28. Ibid.

29. The details of these systems are beyond the scope of this book, so they are not described here. Readers wishing for insight into these techniques should consult Shaler, *Who They Were.*

30. Shaler, *Who They Were.*

31. Ibid., 36.

32. This description of how DNA identification works is based in part on the explanation originally presented in my 2007 book *Genetic Witness.*

33. Shaler, *Who They Were.*

34. Ibid.

35. Ibid., 83.

36. Ibid., 95.

37. Ibid., 94.

38. President's DNA Initiative, *Lessons Learned From 9/11: DNA Identification in Mass Fatality Incidents* (Washington, DC: National Institute of Justice, 2006); Shaler, *Who They Were.*

39. Shaler, *Who They Were.*

40. President's DNA Initiative, *Lessons Learned*; Shaler, *Who They Were.*

41. Shaler, *Who They Were.*

42. Ibid.

43. Eve Conant, "Terror: The Remains of 9/11 Hijackers," *Newsweek*, January 2, 2009, http://www.newsweek.com/terror-remains-911-hijackers-78327.

44. Shaler, *Who They Were.*

45. Office of the Chief Medical Examiner, "World Trade Center Operational Statistics, July 2013."

46. Shaler, *Who They Were, x.*

47. Shaler, *Who They Were.*

48. Shaler, *Who They Were*; Give Your Voice, "Give Your Voice Latest Information," December 10, 2001.

49. Shaler, *Who They Were.*

50. Mike Hennessey, "World Trade Center Identifications: The Administrative Review Process," paper presented at the Proceedings of the Thirteenth International Symposium on Human Identification, 2002.

51. Shaler, *Who They Were*, 148–149.

52. Hennessey, "World Trade Center Identifications."

53. Shaler, *Who They Were*, 145.

54. Give Your Voice, "Meeting Minutes—December 29, 2001 Mayor Giuliani & Give Your Voice Committee," December 29, 2001, http://www.giveyourvoice.bizland.com/december29.html.

55. Give Your Voice, "Give Your Voice Update January 8, 2002," January 8, 2002, http://www.giveyourvoice.bizland.com/january8.html.

56. Give Your Voice, "DNA Information from the Office of the Chief Medical Examiner—Medical Examiners' DNA Help Hotline," March 28, 2002, http://www.giveyourvoice.bizland.com/DNA.html#Hotline.

57. Shaler, *Who They Were.*

58. Give Your Voice, "Minutes of the Meeting with Medical Examiner's Office July 8, 2002," July 8, 2002, http://www.giveyourvoice.bizland.com/ocme78.html.

59. David W. Chen, "Grim Scavenger Hunt for DNA Drags on for Sept. 11 Families," *New York Times*, February 9, 2002, A1.

60. Ibid.

61. Shaler, *Who They Were*, 77.

62. Give Your Voice, "DNA Information from the Office of the Chief Medical Examiner—Why Go Through the Process of Identifying Remains?" (March 28, 2002), http://www.giveyourvoice.bizland.com/DNA.html #Why Go.

63. Shaler, *Who They Were*; Ribowsky and Shachtman, *Dead Center.*

64. N. R. Kleinfeld, "Error Put Body of One Firefighter in Grave of a Firehouse Colleague," *New York Times*, November 28, 2001, A1.

65. Ribowsky and Shachtman, *Dead Center*, 218.

66. Kleinfeld, "Error."

67. Ribowsky and Shachtman, *Dead Center.*

68. Shaler, *Who They Were.*

69. *Huffington Post*, "Diane Horning: NYC Owned Up to Giving Our Family Wrong Remains After 9/11," *HuffPost Live* video, 1.58, April 5, 2013, http://www.huffingtonpost.com/2013/04/05/diane-horning-nyc-gaves-us-remains-911-_n_3023938.html.

70. Diane Horning (mother of WTC victim and founder of WTCFPB), interview with New York State Museum, February 6, 2008, New York State Museum and Archive.

71. Mundorff, interview.

72. Give Your Voice, "DNA Information."

73. Ibid.

74. Ibid.

75. Ibid.

76. Ibid.

77. Mundorff, interview; Ribowsky, interview; Giovanna Vidoli (forensic anthropologist), phone interview with author, January 24, 2014.

78. Vidoli, interview.

79. Ribowsky and Shachtman, *Dead Center,* 194.

80. Gordon Haberman (father of 9/11 victim), phone interview with author, February 11, 2014; Ribowsky and Shachtman, *Dead Center.*

81. Ribowsky and Shachtman, *Dead Center,* 195.

82. Ibid., 194.

83. Amy Waldman, "A Knock on the Door, with the Message of Death," *New York Times*, October 5, 2001, A1.

84. Michelle McPhee, "DNA Gives WTC Cop a Funeral. Hero Buried Today," *New York Daily News*, August 10, 2002, 6.

85. Leslie Casimir and Maki Becker, "Closure, At Last, for Kin of Cop Who Died At WTC," *New York Daily News*, August 11, 2002, 3.

86. Eric Lipton, and James Glanz, "DNA Science Pushed to the Limit in Identifying the Dead of Sept. 11," *New York Times*, April 22, 2002, A1.

87. Ibid.

88. Ribowsky, interview.

89. This policy was groundbreaking at the time, but it is now standard protocol in mass casualty events.

90. Douglas Montero, "Tragic S.I. Family Has Buried Son Four Times," *New York Post*, September 10, 2003, 8.

91. Haberman, interview.

92. Ibid.

93. Anthony Gardner (brother of 9/11 victim), phone interview with author, January 27, 2014.

94. Ibid.

95. Ibid.

96. Ibid.

97. Clemente Lisi, "Still Hoping for the Call—Fireman's Mom Prays He'll be ID'd," *New York Post*, May 30, 2002, 9.

98. Paul H. B. Shin, "As Time Goes By, Hopes Dim on DNA," *New York Daily News*, September 10, 2003, 9.

99. Castrianno, interview; Haberman, interview.

100. Castrianno, interview.

101. Ibid.

102. Shaler, *Who They Were.*

103. Patrice O'Shaughnessy, "More Than Half of Victims IDd," *New York Daily News*, September 11, 2002, 8.

104. Give Your Voice, "WTC DNA Identifications Update 1/25/03," January 25, 2003, http://www.giveyourvoice.bizland.com/index.html#dna130.

105. Ibid.

106. Charles S. Hirsch, "Letter to Family Members Announcing Pause in DNA Identification Efforts," April 22, 2005, http://www.nyc.gov/html/misc/pdf/lettertofamilies.pdf.

107. Quoted in Shaler, *Who They Were,* 321–322.

108. David W. Chen, "As 9/11 Remains Go Unnamed, Families Grieve Anew," *New York Times*, February 24, 2005, B3; Eric Lipton, "At Limits of Science, 9/11 Effort Comes to an End," *New York Times*, April 3, 2005, A29; Adam Lisberg and Paul H. B. Shin, "1,161 Forever Lost to the Fire. Coroner Gives Up Trying to Identify Nearly Half of WTC Victims," *New York Daily News*, February 23, 2005, 5.

109. Charles S. Hirsch, "Letter to World Trade Center Families Announcing Technological Advances by Bode Technologies," September 21, 2006. On file with author.

110. Office of Chief Medical Examiner, "Press Release Re: Testing of Previously Recovered Remains by Bode Technology," June 29, 2007, *Voices of September 11th,* On file with author.

111. Mundorff, interview.

112. Ibid.

113. Ribowsky and Shachtman, *Dead Center*, 237–238.

4: Master Plan

1. Monica Iken (founder of September's Mission), interview with author, July 26, 2011, Manhattan.

2. Ibid.

3. Ibid.

4. Ibid.

5. Deborah Solomon, "From the Rubble, Ideas for Rebirth," *New York Times*, September 30, 2001, 2:37.

6. Ibid.

7. Ibid.

8. Ibid.

9. Public Broadcasting Service, *America Rebuilds: A Year at Ground Zero,* 2002. http://www.pbs.org/americarebuilds.

10. Elizabeth Greenspan, *Battle for Ground Zero: Inside the Political Struggle to Rebuild the World Trade Center* (New York: Palgrave Macmillan, 2013).

11. Paul Goldberger, *Up from Zero: Politics, Architecture, and the Rebuilding of New York* (New York: Random House, 2004).

12. Greenspan, *Battle for Ground Zero.*

13. Goldberger, *Up from Zero.*

14. Michael Cooper, "Official Says Bush Renewed Pledge of $20 Billion in Aid," *New York Times,* January 25, 2002, B6; Lower Manhattan Development Corporation, "HUD Funding," n.d., http://www.renewnyc.com/fundinginitiatives/; Edward Wyatt, David W. Chen, Charles V. Bagli, and Raymond Hernandez, "After 9/11, Parcels of Money, and Dismay," *New York Times,* December 30, 2002, A1.

15. Goldberger, *Up from Zero*; Greenspan, *Battle for Ground Zero.*

16. Goldberger, *Up from Zero.*

17. Lower Manhattan Development Corporation, "Governor and Mayor Name Lower Manhattan Redevelopment Corporation," November 29, 2001, http://www.renewnyc.com/displaynews.aspx?newsid=2b0bfde6-61b6-48f9-a6ca-70475846c95b.

18. Ibid.

19. Robert Ivy, "The Protetch Show," January 18, 2002, https://web.archive.org/web/20140803211013/http://archrecord.construction.com/news/fromTheField/archives/020118protetch.asp.

20. Max Protetch, *A New World Trade Center: Design Proposals from Leading Architects Worldwide* (New York: HarperCollins, 2002).

21. Manhattan Community Board No. 1, "About the District," n.d., http://www.nyc.gov/html/mancb1/html/district/about.shtml. Accessed January 11, 2016.

22. Madelyn Wils (CB 1 board director), interview, 2002, *America Rebuilds: A Year at Ground Zero,* PBS, http://www.pbs.org/americarebuilds/profiles/profiles_wils_t.html.

23. Public Broadcasting Service, *America Rebuilds.*

24. Iken, interview, *America Rebuilds.*

25. Public Broadcasting Service, *America Rebuilds.*

26. Civic Alliance to Rebuild Downtown New York, "Listening to the City: Report of Proceedings," February 7, 2002, 2, http://www.icisnyu.org/admin/files/ListeningtoCity.pdf.

27. Ibid., 9.

28. Ibid.

29. Ibid.

30. Ibid., 2.

31. Ibid., 6.

32. Ibid., 8.

33. Ibid., 7.

34. Ibid.

35. Louis R. Tomson, "Opening Remarks of President Louis R. Tomson, Federal Hall, New York," July 16, 2002, http://renewnyc.com/content/speeches/presidents_remarks2002_07_16.pdf.

36. Lower Manhattan Development Corporation, "About Us," n.d., http://www.renewnyc.com/AboutUs/AdvisoryMeetings.aspx. Acessed January 11, 2016. The original twenty-four members of the FAC were Virginia Bauer, Paula Grant Berry, Darlene Dwyer (Windows of Hope), Kathy Ashton (Give Your Voice), Christy Ferer (Mayor's Family Liaison), Mary Fetchet (Voices of September 11th), Marian Fontana (September 11th WVFA), Anthony Gardner (WTC United Family Group), Elinore Hartz, Monica Iken (September's Mission), Lee Ielpi (September 11th WVFA), Ann Johnson, Anthoula Katsimatides, Carie Lemack (Families of September 11), Edie Lutnick (Cantor Fitzgerald Relief Fund), Michael Macko, Kathleen Lynch Martens, Maria McHugh, Vincent F. Pitta, Sally Regenhard (Campaign for Skyscraper Safety), Tom Rogér (Families of September 11), William Rodriguez (Hispanic Victims Group), Jim Smith, and Nikki Stern.

37. Iken, interview, *America Rebuilds.*

38. Coalition of 9/11 Families, *Tribute* 1, no. 1 (July 2002): 1, http://www.911families.org/wp-content/uploads/2002_07.pdf. All coalition newsletters available at Newsletter Archive, September 11th Families' Association, http://www.911families.org/about-us/newsletter-archive/.

39. Ibid., 3.

40. Lee Ielpi, "Memorial Complex," *Tribute* (Coalition of 9/11 Families) 2, no. 1 (2003): 1.

41. Coalition of 9/11 Families, *Tribute*, 3.

42. Ibid.

43. Civic Alliance to Rebuild Downtown New York, "Listening to the City: Report of Proceedings" (September 2002), 9. https://web.archive.org/web/20040726142406/http://www.civic-alliance.org/pdf/0920FinalLTCReport.pdf

44. "Proposals for Downtown Draw Array of Opinions," *New York Times*, July 21, 2002, http://www.nytimes.com/2002/07/21/nyregion/21VOIC.html.

45. Coalition of 9/11 Families, *Tribute* 1, no. 1. This sentiment was expressed to me in numerous family interviews as well.

46. Edie Lutnick, *An Unbroken Bond: The Untold Story of How the 658 Cantor Fitzgerald Families Faced the Tragedy of 9/11 and Beyond* (New York: Emergence Press, 2011).

47. Ibid., 97.

48. Ibid., 97–98.

49. Goldberger, *Up from Zero.*

50. Ibid.

51. Ibid.

52. Ibid.

53. A slideshow of the various designs can be found at http://www.pbs.org/wgbh/pages/frontline/shows/sacred/designs/prelim.html.

54. Robert Campbell, "Shell Game Comes Up Office Ghetto Every Time It's Played," *Boston Globe*, July 17, 2002; Ada Louise Huxtable, "Another World Trade Center Horror," *Wall Street Journal*, July 25, 2002, B13.

55. Coalition of 9/11 Families, *Tribute* 1, no. 1.

56. Iken, interview, *America Rebuilds.*

57. Ibid.; Ielpi, "Memorial Complex."

58. Iken, interview, *America Rebuilds.*

59. Edward Wyatt, "Pataki's Surprising Limit on Ground Zero Design," *New York Times*, July 2, 2002, B1.

60. Glenn Collins, and David W. Dunlap, "Fighting for the Footprints of Sept. 11; Families Renew Call for a Memorial That Includes Traces of Towers Themselves," *New York Times*, December 30, 2003, B1; Iken, interview, *America Rebuilds.*

61. Lutnick, *An Unbroken Bond.*

62. Wyatt, "Pataki's Surprising Limit."

63. Civic Alliance, "Listening to the City: Report of Proceedings."

64. Ibid., 2.

65. Ibid., 10.

66. Ibid.

67. Ibid.

68. See Ibid., 6 for participant demographics.

69. Goldberger, *Up from Zero.*

70. Greenspan, *Battle for Ground Zero.*

71. Liam Strain, reprinted in "Proposals for Downtown Draw Array of Opinions," *New York Times*, July 21, 2002, A30.

72. Goldberger, *Up from Zero.*

73. Lower Manhattan Development Corporation, "Lower Manhattan Development Corporation Announces Design Study for World Trade Center Site and Surrounding Areas," August 14, 2002, http://www.renewnyc.com/displaynews.aspx?newsid=da800006-c35b-4f1c-a9ec-ff53cfe45ae2.

74. Goldberger, *Up From Zero.*

75. Philip Nobel, *Sixteen Acres: Architecture and the Outrageous Struggle for the Future of Ground Zero* (New York: Metropolitan Books, 2005).

76. Michael Sorkin, *Starting from Zero: Reconstructing Downtown New York* (New York: Routledge, 2003), 58.

77. Nobel, *Sixteen Acres*, 96.

78. Civic Alliance to Rebuild Downtown New York, "Listening to the City: Report of Proceedings," September 2002. http://www.icisnyu.org/admin/files/ListeningtoCity.pdf.

79. New York New Visions, "Principles for the Rebuilding of Lower Manhattan" (February 2002), 3, https://web.archive.org/web/20020124232546/http://nynv.aiga.org/nynv_book.pdf.

80. Lower Manhattan Development Corporation, "Request for Qualifications: Innovative Designs for the World Trade Center Site," August 19, 2002, 2, http://www.renewnyc.com/content/rfps/RFQInnovativeDesignStudy.pdf.

81. Lower Manhattan Development Corporation, "Lower Manhattan Development Corporation Announces Design Study for World Trade Center Site," August 14, 2002, http://www.prnewswire.com/news-releases/lower-manhattan-development-corporation-announces-design-study-for-world-trade-center-site-and-surrounding-areas-76711737.html. A request for qualifications is not a solicitation for designs—it is more like a job application in which interested architects make the case for being hired to work with the client on the design of a building or development.

82. Nobel, *Sixteen Acres,* 106.

83. Lower Manhattan Development Corporation, "Lower Manhattan Development Corporation Announces Six Teams of Architects and Planners to Participate in Design Study of World Trade Center Site," September 26, 2002, http://www.renewnyc.com/displaynews.aspx?newsid=655798f4-b8ee-4583-bf27-b3b9a9c29f1f.

84. Goldberger, *Up from Zero.*

85. Studio Daniel Libeskind, "Introduction," 2002, Lower Manhattan Development Corporation, http://www.renewnyc.com/plan_des_dev/wtc_site/new_design_plans/firm_d/default.asp.

86. Ibid.

87. Ibid.

88. Foster and Partners, "Introduction," 2002, Lower Manhattan Development Corporation, http://www.renewnyc.com/plan_des_dev/wtc_site/new_design_plans/firm_a/default.asp.

89. Richard Meier & Partners, et al., "Introduction," Lower Manhattan Development Corporation, 2002, http://www.renewnyc.com/plan_des_dev/wtc_site/new_design_plans/firm_g/ default.asp.

90. Think Design, "Introduction," Lower Manhattan Development Corporation, 2002, http://www.renewnyc.com/plan_des_dev/wtc_site/new_design_plans/firm_e/default.asp.

91. United Architects, "Introduction," Lower Manhattan Development Corporation, 2002, http://www.renewnyc.com/plan_des_dev/wtc_site/new_design_plans/firm_f/default.asp.

92. SOM et al., "Introduction," Lower Manhattan Development Corporation, 2002, http://www.renewnyc.com/plan_des_dev/wtc_site/new_design_plans/firm_c/default.asp.

93. Peterson/Littenberg Architecture and Urban Design, "Introduction," Lower Manhattan Development Corporation, 2002, http://www.renewnyc.com/plan_des_dev/wtc_site/new_design_plans/firm_b/default.asp.

94. Lower Manhattan Development Corporation, "Proposed Designs for World Trade Center Site Released at the Winter Garden," December 18, 2002, http://www.renewnyc.com/displaynews.aspx?newsid=6ec82517-cd24-422e-86a7-104ea4446dee.

95. Lower Manhattan Development Corporation, "LMDC and Port Authority Narrow Field of Design Concepts under Consideration for World Trade Center Site," February 4, 2003, http://www.renewnyc.com/displaynews.aspx?newsid=ef439d25-9e94-4e38-bc34-8cea4f3262a5.

96. Bruce DeCell, "LMDC Design Proposals," *Tribute* (Coalition of 9/11 Families) 2, no. 1 (2003): 3.

97. New York New Visions, "Evaluation of Innovative Design Proposals," January 2003, 4. http://nynv.aiga.org/NYNV20030113.pdf.

98. Ibid., 5.

99. Ibid., 2.

100. Lower Manhattan Development Corporation, "LMDC and Port Authority Narrow Field."

101. Edward Wyatt, "Ground Zero Plan Seems to Circle Back," *New York Times*, September 13, 2003, B1.

102. Alexander Garvin, "The Rebuilding of a City," *Perspecta* 36 (2005): 92–99.

103. Goldberger, *Up from Zero*; Greenspan, *Battle for Ground Zero*; Nobel, *Sixteen Acres*.

104. Lower Manhattan Development Corporation, "The Lower Manhattan Development Corporation and Port Authority of New York & New Jersey Announce Selection of Studio Daniel Libeskind: Memory Foundations as Design Concept for World Trade Center Site," February 27, 2003, http://www.renewnyc.com/displaynews.aspx?newsid=41c07ff1-9b1a-41a2-866b-8aa8148b6736.

105. Beverly Eckert, "LMDC Modifies Libeskind Plan," *Tribute* (Coalition of 9/11 Families) 2, no. 2 (2003): 1.

106. Ibid.

5: Memorial

1. Drew Gilpin Faust, *This Republic of Suffering: Death and the American Civil War* (New York: Knopf, 2008), xiv.

2. Ibid.

3. Ward Churchill, " 'Some People Push Back': On the Justice of Roosting Chickens," September 2001, http://www.kersplebedeb.com/mystuff/s11/churchill.html.

4. Margaret Talbot, "The Lives They Lived: 3,225 (At Last Count)," *New York Times Magazine*, December 30, 2001, 16; "World Trade Center Site Memorial

Competition Jury Statement," November 19, 2003, *New York Times,* http://www.nytimes.com/2003/11/19/nyregion/19WTC-JURY-TEXT.html; James E. Young, "The Memorial Process: A Juror's Report from Ground Zero," in *Contentious City: The Politics of Recovery in New York City,* ed. John Mollenkopf (New York: Russell Sage Foundation, 2005), 140–162.

5. Amy Waldman, "Posters of the Missing Now Speak of Losses," *New York Times,* September 29, 2001, B1.

6. Ibid.

7. Immediate memorials have become a predictable response to sudden public deaths. Harriet F. Senie, *Memorials to Shattered Myths: Vietnam to 9/11* (New York: Oxford University Press, 2016).

8. Wendy Carlos, "Aftermath September 11, 2001: Images and Words from the Trenches," 2001, http://www.wendycarlos.com/aftermath/amath.html.

9. Siena College, "9/11 Exhibit Stirs Emotion and Remembrance," September 8, 2011, *Siena News,* 13, https://issuu.com/sienacollege/docs/sienanewsfall2011.

10. Carlos, "Aftermath."

11. Harriet F. Senie, "A Difference in Kind: Spontaneous Memorials After 9/11," *Sculpture Magazine* (July/August 2003), http://www.sculpture.org/documents/scmag03/jul_aug03/webspecial/senie.shtml

12. Ibid.

13. Diane Horning (mother of WTC victim and founder of WTCFPB), interview with New York State Museum, February 6, 2008, New York State Museum and Archive; Sally Regenhard (founder of Skyscraper Safety Campaign), interview with author, July 24, 2011, World Trade Center site.

14. New York New Visions, "Memorials Process Team Briefing Book: Findings From the Outreach, Temporary Memorials & Research Working Groups" (March 2002), https://web.archive.org/web/20120313114925/http://nynv.aiga.org/pdfs/NYNV_MemorialsBriefingBook.pdf.

15. Ibid.

16. Ibid.

17. Ibid.

18. Ibid.

19. Ibid.

20. Ibid.

21. Julie V. Iovine, "Memorials Proliferate in Crowded Downtown," *New York Times,* March 13, 2003, E1.

22. Madelyn Wils (CB1 board chair), interview, 2002, *America Rebuilds: A Year at Ground Zero,* PBS, http://www.pbs.org/americarebuilds/profiles/profiles_wils_t.html.

23. Iovine, "Memorials Proliferate."

24. Monica Iken (founder of September's Mission), interview, 2002, *America Rebuilds: A Year at Ground Zero,* PBS, http://www.pbs.org/americarebuilds/profiles/profiles_iken_t.html.

25. Ibid.

26. Ibid.

27. Jennifer Senior, "The Memorial Warriors," September 15, 2002, http://nymag.com/nymetro/news/sept11/features/n_7691/.

28. Ibid.

29. Ibid.

30. Nikki Stern, *Because I Say So: The Dangerous Appeal of Moral Authority* (Minneapolis: Bascom Hill Books, 2010).

31. Lower Manhattan Development Corporation, minutes, Program Sub-Committee of the Families Advisory Council, June 4, 2002, http://www.renewnyc.com/content/meetings/06.04.02Sub-CommitteeMemorialMinutesFINAL.pdf.

32. Lower Manhattan Development Corporation, "The Public Dialogue Draft Memorial Mission Statement and Program for the World Trade Center Memorial," March 12, 2003, http://www.renewnyc.com/content/pdfs/public_dialogue_memorial_03-03.pdf.

33. Lower Manhattan Development Corporation, "Committees Created to Draft WTC Memorial Mission Statement and Program," November 12, 2002, http://www.renewnyc.com/displaynews.aspx?newsid=e3f87188-1ed5-4193-943b-5fd10befab20.

34. Lower Manhattan Development Corporation, "Memorial Mission Statement and Memorial Program," 2003, http://www.renewnyc.com/memorial/memmission.asp.

35. Edward Wyatt, "Victims' Families Sense Influence on Ground Zero Plans Is Waning," *New York Times*, November 16, 2002, A1.

36. Edie Lutnick, *An Unbroken Bond: The Untold Story of How the 658 Cantor Fitzgerald Families Faced the Tragedy of 9/11 and Beyond* (New York: Emergence Press, 2011).

37. Edward Wyatt, "Outspoken 9/11 Widow Joins Memorial Panel," *New York Times*, November 28, 2002, B4.

38. Ibid.

39. Lower Manhattan Development Corporation, "Memorial Research Tour: Shanksville, PA, Washington, DC, Oklahoma City, OK, Montgomery, AL," October 2002, www.renewnyc.com/memorial/2002MemorialTripPresentation.ppt.

40. Ibid., slide 25.

41. Ibid.

42. New York New Visions, "Four Principles for the Memorial Program," December 16, 2002, https://web.archive.org/web/20120312163256/http://nynv.aiga.org/FamResReptFinal.pdf.

43. Ibid.

44. Compare: Lower Manhattan Development Corporation, "Family Outreach Letter," Summer/Fall 2002, http://www.renewnyc.com/content/pdfs/Outreach-August2002.pdf; Lower Manhattan Development Corporation, "Family Outreach Letter," January 6, 2003, http://www.renewnyc.com/content/pdfs/Outreach

-January2003.pdf; and Lower Manhattan Development Corporation, "World Trade Center Site: Memorial Competition Guidelines," 2003, https://www.911memorial.org/sites/all/files/LMDC%20Memorial%20Guidelines.pdf.

45. Lower Manhattan Development Corporation, "Family Outreach Letter," January 6, 2003.

46. Lower Manhattan Development Corporation, "Public Dialogue."

47. Lower Manhattan Development Corporation, "Memorial Competition Guidelines," 18–19.

48. Ibid.

49. Ibid., 19.

50. Ibid., 10.

51. Young, "The Memorial Process."

52. Ibid.

53. Lower Manhattan Development Corporation, "Memorial Competition Guidelines," 10.

54. Lower Manhattan Development Corporation, "Meeting Summary, Families Advisory Council," October 16, 2003, http://www.renewnyc.com/content/meetings/FAC_Meeting_Summary_10-16-03.pdf.

55. Paul Goldberger, *Up From Zero: Politics, Architecture, and the Rebuilding of New York* (New York: Random House, 2004).

56. Lower Manhattan Development Corporation, "Memorial Competition Guidelines," 26.

57. Goldberger, *Up From Zero.*

58. Ibid., 219.

59. Ibid.

60. Young, "The Memorial Process," 148.

61. Lower Manhattan Development Corporation, "World Trade Center Site Memorial Competition: Jury Members," 2003, http://www.wtcsitememorial.org/about_jury.html.

62. Lutnick, *An Unbroken Bond.*

63. Ibid.

64. Edward T. Linenthal, *The Unfinished Bombing: Oklahoma City in American Memory* (New York: Oxford University Press, 2001).

65. Ibid., 204.

66. Lower Manhattan Development Corporation, "Wall Street Rising Breakfast Creating a Memorial at the World Trade Center Site: Next Steps," April 8, 2003, http://www.renewnyc.com/content/speeches/04.07.03.pdf.

67. Lower Manhattan Development Corporation, "Joint Meeting of Memorial Competition Jury and Family Advisory Council," May 27, 2003, 6, http://www.renewnyc.com/content/pdfs/factranscript2.pdf.

68. Ibid., 16.

69. Ibid., 40.

70. Ibid., 33.

71. Ibid., 33–34.

72. Ibid., 34.

73. Ibid., 36.

74. Ibid., 47.

75. Ibid., 48.

76. Ibid., 44.

77. Lower Manhattan Development Corporation, "Joint Meeting of Memorial Competition Jury -and- All Advisory Councils," June 5, 2003, http://www.renewnyc.com/content/pdfs/alladvisorycounciltranscript.pdf.

78. Ibid., 58.

79. Ibid., 26–27.

80. Ibid., 28.

81. Ibid.

82. Ibid, 29.

83. Ibid., 40; Maggie Haberman, "Panel to Decide on a Separate 9/11 Memorial," *New York Daily News*, May 31, 2003, 6.

84. Advocates for a 9/11 Fallen Heroes Memorial, "Our Mission" n.d., https://web.archive.org/web/20150309221402/http://911fallenheroes.org/ourmission.html.

85. Josh Rogers, "Thousands Vie to Design W.T.C. Memorial," *Downtown Express*, June 2003, http://www.downtownexpress.com/DE_06/thousandsvie.html.

86. Allison Blais, and Lynn Rasic, *A Place of Remembrance: Official Book of the National September 11 Memorial* (Washington, DC: National Geographic Society, 2011).

87. Rogers, "Thousands Vie."

88. Blais and Rasic, *A Place of Remembrance.*

89. Young, "The Memorial Process."

90. Goldberger, *Up From Zero.*

91. Young, "The Memorial Process," 154.

92. All finalists can be viewed at: http://www.wtcsitememorial.org/finalists.html.

93. Ibid.

94. "World Trade Center Site Memorial Competition Jury Statement."

95. Ibid.

96. Herbert Muschamp, "Amid Embellishment and Message, a Voice of Simplicity Cries to be Heard," *New York Times*, November 20, 2003, B3.

97. Glenn Collins, "8 Designs Confront Many Agendas at Ground Zero," *New York Times*, November 20, 2003, A1.

98. Municipal Arts Society, "Imagine New York III: Toward the People's Memorial," December 2003, http://www.gothamgazette.com/images/pdf/rebuildingnyc/imaginereport.pdf.

99. Ibid.

100. New York New Visions, letter to the Memorial Competition Jury, November 26, 2003, http://www.nyplanning.org/docs/nynvfinalmem12-1-03.pdf.

101. Ibid.

102. Goldberger, *Up From Zero.*

103. Harriet F. Senie, "Absence Versus Presence: The 9/11 Memorial Design," *Sculpture Magazine,* 23, no. 4 (2004), http://www.sculpture.org/documents/scmag04/may04/forum/forum.shtml.

104. Michael Kimmelman, "Ground Zero's Only Hope: Elitism," *New York Times,* December 7, 2003, B1.

105. Goldberger, *Up From Zero.*

106. Young, "The Memorial Process," 157.

107. Ibid., 156.

108. Ibid.

109. Goldberger, *Up from Zero;* Joe Hagan, "The Breaking of Michael Arad," *New York Magazine,* May 22, 2006, 20–27, 100–102.

110. Young, "The Memorial Process."

111. Ibid.

112. Ibid.

113. Michael Arad and Peter Walker, "Reflecting Absence," World Trade Center Site Memorial Competition, 2003, http://www.wtcsitememorial.org/fin7.html.

114. Ibid.

115. Hagan, "The Breaking of Michael Arad."

116. Ibid.

6: Remaking the Memorial

1. Elizabeth Greenspan, *Battle for Ground Zero: Inside the Political Struggle to Rebuild the World Trade Center* (New York: Palgrave Macmillan, 2013), 148.

2. David W. Dunlap, "1,776-Foot Design Is Unveiled for World Trade Center Tower," *New York Times,* December 20, 2003, A1.

3. Greenspan, *Battle for Ground Zero.*

4. Ibid.

5. Joe Hagan, "The Breaking of Michael Arad," *New York Magazine,* May 22, 2006, 20–27, 100–102.

6. Ibid.

7. Ibid.

8. Lower Manhattan Development Corporation, "Minutes, Meeting of the Directors," April 13, 2004, http://www.renewnyc.com/content/meetings/April_minutes_final.pdf.

9. Lynne Duke, "Blueprint of a Life: Architect J. Max Bond Jr. Has Had to Build Bridges to Reach Ground Zero," *Washington Post,* July 1, 2004, C1.

10. Goldberger, *Up From Zero*; Hagan, "The Breaking of Michael Arad"; Robin Pogrebin, "Pushed and Pulled, Designer of 9/11 Memorial Focuses on the Goal," *New York Times*, May 10, 2005, E1.

11. Hagan, "The Breaking of Michael Arad."

12. Lower Manhattan Development Corporation, "Progress Report, 2001–2004," 2004, http://www.renewnyc.com/content/pdfs/annual_report_Nov2004.pdf.

13. International Freedom Center, "International Freedom Center Fact Sheet," http://www.renewnyc.org/content/pdfs/IFC_Fact_Sheet.pdf. Accessed January 11, 2016.

14. Ibid.

15. Debra Burlingame, "The Great Ground Zero Heist," *Wall Street Journal*, June 7, 2005, A14.

16. Ibid.

17. Take Back the Memorial, "About," July 1, 2005, http://www.takebackthememorial.net/about.htm.

18. Edie Lutnick. *An Unbroken Bond: The Untold Story of How the 658 Cantor Fitzgerald Families Faced the Tragedy of 9/11 and Beyond* (New York: Emergence Press, 2011), 206.

19. Burlingame, "Great Ground Zero Heist."

20. Robert Kolker, "The Grief Police," *New York Magazine*, November 28, 2005, 46–56, http://nymag.com/nymetro/news/sept11/features/15140.

21. Ibid.

22. Patrick D. Healy, "Pataki Warns Cultural Groups for Museum at Ground Zero," *New York Times*, June 25, 2005, B1.

23. Kolker, "Grief Police."

24. Put It Above Ground, "Mission," https://wayback.archive-it.org/1029/20080306015757/http://www.putitaboveground.org/mission/. Accessed January 11, 2016.

25. Ibid.

26. Dennis McKeon, "Who is Rosaleen Tallon and Why is She Sleeping on the Street," *Where to Turn*, March 16, 2006, http://www.where-to-turn.org/phpBB2/viewtopic.php?t=18944&sid=7b858d6a98601a1fe619e2554d49d106.

27. Dennis McKeon, "Petition Update Day 99 14,134 Signatures," *Where to Turn*, June 14, 2006, http://www.where-to-turn.org/phpBB2/viewtopic.php?t=24136&sid=9d19d461ed52753dcd94315f957d8993.

28. McKeon, "Who is Rosaleen Tallon."

29. See, for example, Laura Trevelyan, "Families Divided over NY Memorial," *BBC News*, March 13, 2006, http://news.bbc.co.uk/2/hi/americas/4802772.stm.

30. David W. Dunlap, "Cost and Safety Put Memorial's Striking Vision at Risk," *New York Times*, May 11, 2006, B2.

31. David W. Dunlap, "Contractor Sought for Part 1 of Ground Zero Memorial," *New York Times*, January 3, 2006, B3.

32. The construction documents were removed from LMDC's website after the symbolic mortuary vessel plan was abandoned.

33. Ibid.

34. Ibid. See also Matter of Coalition of 9/11 Families Inc. vs. Lower Manhattan Development Corporation, 12 Misc 3d 1173(A) (Supr. Ct. of NY County, 2006).

35. Ibid.

36. 9/11 Memorial, "Summary of Outreach Regarding Plans to Relocate the City of New York's Office of Chief Medical Examiner's (OCME's) Repository for the Unidentified and Unclaimed Remains of 9/11 Victims," n.d., https://s3.amazonaws.com/s3.documentcloud.org/documents/97992/memorial-museum-response.pdf.

37. David W. Dunlap, "Discord Over 9/11 Memorial's Symbolism," *New York Times*, January 12, 2006, B3.

38. Lower Manhattan Development Corporation, "Memorial Competition Guidelines," 2003, 19, https://www.911memorial.org/sites/all/files/LMDC%20Memorial%20Guidelines.pdf.

39. Coalition of 9/11 Families, "United Vision of Remembrance for Future Generations," *Tribute* 2, no. 2 (July 2003): 3, http://www.911families.org/wp-content/uploads/2003_07.pdf.

40. Ibid.

41. 9/11 Memorial, "Summary of Outreach."

42. Trevelyan, "Families Divided"; Matter of Coalition of 9/11.

43. Matter of Coalition of 9/11, at p. 11 of opinion.

44. Michael Arad and Peter Walker, "Reflecting Absence," World Trade Center Site Memorial Competition, 2003, http://www.wtcsitememorial.org/fin7.html.

45. Lutnick, *An Unbroken Bond*, 195.

46. Ibid.

47. Ibid.

48. Ibid., 185.

49. "Naming Names at Ground Zero," *New York Times*, July 5, 2006, A16.

50. Ibid.

51. Hagan, "The Breaking of Michael Arad."

52. Frank J. Sciame, "World Trade Center Memorial Draft Recommendations and Analysis," Lower Manhattan Development Corporation, Renew New York City, June 15, 2006, http://www.renewnyc.com/content/pdfs/FRANK_SCIAME_Draft_Rec_Analysis.pdf.

53. Hagan, "The Breaking of Michael Arad."

54. Sciame, "World Trade Center Memorial Draft Recommendations."

55. Ibid.

56. Lower Manhattan Development Corporation, "Meeting of the Families Advisory Council (With Frank J. Sciame)," May 31, 2006, 10. From the collection of the Lower Manhattan Development Corporation. On file with author.

57. Ibid., 11–12.

58. Ibid., 21.

59. Ibid.

60. Ibid., 28.

61. Ibid., 31.

62. Ibid., 32.

63. Ibid., 42.

64. Ibid., 44.

65. Ibid.

66. Ibid., 104.

67. Dunlap, "Cost and Safety."

68. Lower Manhattan Development Corporation, "Meeting of the Families Advisory Council," 60.

69. Ibid., 78.

70. Ibid., 108–109.

71. Ibid., 117.

72. See, for example, Steve Cuozzo, "Rebuilding Is the Only Reason for the Find," *New York Post*, October 24, 2006, 16; "Naming Names at Ground Zero;" Kolker, "Grief Police."

73. Brooklyn Rider, "Comment in Thread: WTC Memorial—By Michael Arad (Architect) and Peter Walker (Landscape)," *Wired New York* (blog), August 8, 2005, http://wirednewyork.com/forum/showthread.php?t=4463&page=12.

74. Sciame, "World Trade Center Memorial Draft Recommendations," 11.

75. Marita Sturken, *Tourists of History: Memory, Kitsch, and Consumerism from Oklahoma City to Ground Zero* (Durham, NC: Duke University Press, 2007).

76. Sewell Chan, "A New, Longer Name for the 9/11 Memorial," *City Room* (blog), *New York Times*, August 14, 2007, http://cityroom.blogs.nytimes.com/2007/08/14/a-new-longer-name-for-the-911-memorial/.

77. Lower Manhattan Development Corporation, "Presentation to Section 106 Consulting Parties," September 29, 2006, http://www.renewnyc.com/content/pdfs/060929-section%20106.pdf.

78. Sara Kugler, "Mayor Bloomberg Takes Over WTC Memorial Despite Earlier Criticism of Project," Associated Press, October 7, 2006.

79. Lutnick, *An Unbroken Bond.*

80. David W. Dunlap, "Plan is Changed for Arranging Names on Trade Center Memorial," *New York Times*, December 14, 2006, B3.

81. Lutnick, *An Unbroken Bond,* 251.

82. Amy Westfeldt, "Bloomberg: $300m Raised for WTC Memorial," Associated Press, April 17, 2007. He ultimately only left office at the end of 2013. The city council granted him an extension on term limits so that he could handle the city's response to the 2008 financial crisis.

83. Ibid.

84. Dennis McKeon, "Foundation Cracks," letter to the editor, *Downtown Express,* October 13, 2006, http://www.downtownexpress.com/de_179/letterstotheeditor.html.

85. Kugler, "Mayor Bloomberg Takes Over."

86. Put It Above Ground, "Make Sure It's History, Not 'His Story,'" November 24, 2006, http://www.where-to-turn.org/phpBB2/viewtopic.php?t=33177; Put It Above Ground, "This is Not About the Victims of 9/11," February 23, 2007, http://www.where-to-turn.org/phpBB2/viewtopic.php?t=35661.

87. Anthony Gardner (brother of 9/11 victim), phone interview with author, January 27, 2014.

88. Port Authority of New York and New Jersey, "World Trade Center Report: A Roadmap Forward," October 2, 2008, 25, http://www.panynj.gov/wtcprogress/pdf/wtc_report_oct_08.pdf.

89. Ibid., 17.

90. Ibid., 4.

91. Ibid., 3.

92. Ibid.

93. Ibid.

94. Ibid., 5.

95. Ibid.

7: New Finds

1. Eric Lipton and James Glanz, "Victims' Remains Found Near Ground Zero," *New York Times*, June 8, 2002, B1.

2. Richard Weir, "Grim Discovery: Bones from WTC Found on Nearby Bldg," *New York Daily News*, September 10, 2003, 9.

3. David W. Dunlap, "Bovis Is Awarded Deal to Demolish a Tainted Tower at Ground Zero," *New York Times*, August 12, 2005, B2.

4. Ibid.; Lower Manhattan Development Corporation, "Frequently Asked Questions: The Deutsche Bank Building at 130 Liberty Street," n.d., http://www.renewnyc.com/plan_des_dev/130liberty/faq_130liberty_findings.asp.

5. Greg B. Smith, "More 9-11 Pain Unearthed. 2 Whole Buildings Never Searched at Site's Edge," *New York Daily News*, October 22, 2006, 6.

6. David W. Dunlap, "Bone Fragments Found on a Roof near Ground Zero," *New York Times*, September 27, 2005, B1.

7. David W. Dunlap, "Remains Found on Skyscraper Near Ground Zero Are Human," *New York Times*, October 28, 2005, B5.

8. World Trade Center Families for a Proper Burial, "Media Statement from WTC Families for a Proper Burial," *Where to Turn* (blog), September 29, 2005, http://where-to-turn.org/phpBB2/viewtopic.php?t=8383&sid=29f5cfe040d3ea14398cb2b4001ccb73.

9. Giovanna Vidoli, letter to David Ridley, project manager for 130 Liberty Street cleanup, September 26, 2005, http://www.renewnyc.com/content/pdfs/130liberty/protocol-1.pdf.

10. Sara Kugler, "As New Bones Are Found Near WTC Site, Many Families Have No Remains," Associated Press, September 8, 2006.

11. Jim Dwyer, "Pieces of Bone Are Found on Building at 9/11 Site," *New York Times*, April 6, 2006, B4; Diane Horning (mother of WTC victim and founder of WTCFPB), interview with New York State Museum, February 6, 2008, New York State Museum and Archive; Vidoli, letter to David Ridley.

12. Diane Horning, Rosaleen Tallon, Sally Regenhard, Dennis McKeon, and Jim McCaffrey, "9/11 Families Call on NYC Mayor's Office to Oversee Retrieval of Human Remains," *Where to Turn* (blog), April 8, 2006, http://www.where-to-turn.org/phpBB2/viewtopic.php?t=20282.

13. Ibid.

14. David W. Dunlap, "New Concerns about Razing of Bank Tower at Ground Zero," *New York Times*, April 14, 2006, B3.

15. This sentiment was repeated by several people I interviewed.

16. Skyscraper Safety Campaign, "Skyscraper Safety Campaign Responds to the 300 Human Remains Found at Ground Zero," *Where to Turn* (blog), April 14, 2006, http://www.where-to-turn.org/phpBB2/viewtopic.php?t=20651.

17. Vidoli, letter to David Ridley.

18. Skyscraper Safety Campaign, "Skyscraper Safety Campaign."

19. Sally Regenhard, Al Regenhard, Dennis McKeon, Jim McCaffrey, and Rosaleen Tallon, "Letter to Eliot Spitzer Re: Human Remains," *Where to Turn* (blog), July 27, 2006, http://www.where-to-turn.org/phpBB2/viewtopic.php?t=26535&sid=306e5380851cc95dfcca7ca10de72ec1.

20. Patrick Brennan, "Mayor's Response to Our Letter Re: 130 Liberty," *Where to Turn* (blog), July 6, 2006, http://www.where-to-turn.org/phpBB2/viewtopic.php?t=25279.

21. Michael Goodwin, "No End to Cruel Postscript," *New York Daily News*, April 19, 2006, 31.

22. Anthoula Katsimatides, letter to family members, April 14, 2006, http://www.where-to-turn.org/phpBB2/viewtopic.php?t=20635.

23. See, for example, Greg B. Smith, "A Grave Oversight. 9-11 Kin Wait While Much of Deutsche Still Not Searched," *New York Daily News*, April 30, 2006, 8.

24. Greg B. Smith, "Don't Bank on Anything. Search for Remains Stalls as Safety Clouds Deutsche's Fate," *New York Daily News*, May 14, 2006, 11.

25. According to Amy Mundorff, the delay occurred at least in part because of the overwhelming size of the job, and the need to ensure that certain areas were safe before they searched, not negligence or malfeasance. Mundorff, e-mail to author, November 17, 2014.

26. Ibid.

27. Richard A. Gould and Ann Marie Mires, letter to Mayor Bloomberg, July 4, 2006, private collection of Richard A. Gould.

28. Ibid.

29. Kareem Fahim and Anthony DePalma, "9/11 Victims' Kin Assail Search for Remains," *New York Times*, July 15, 2006, B2.

30. Larry McKee, letter to Richard A. Gould, July 16, 2006, private collection of Richard A. Gould.

31. Richard A. Gould, e-mail to Larry McKee, July 17, 2006, private collection of Richard A. Gould.

32. Victor J. Gallo, letter to Charles S. Hirsch, July 25, 2006, private collection of Richard A. Gould.

33. Richard A. Gould, "Final Report of Archaeological Site Visit to the Deutsche Bank Building, New York, August 8, 2006," August 25, 2006, 6, private collection of Richard A. Gould.

34. Richard A. Gould, letter/e-mail to Ellen Borakove, August 2006, private collection of Richard A. Gould.

35. Gould, "Final Report," 6.

36. Ibid.

37. David W. Dunlap, "Expert Supports Search Methods for 9/11 Remains at Bank Building," *New York Times*, August 26, 2006, B2.

38. Greg B. Smith, "No Remains in Bank Bldg., 9-11 Kin Told," *New York Daily News*, September 29, 2006, 33.

39. Al Regenhard and Sally Regenhard, letter to Ruth Simmons, September 11, 2006, private collection of Richard A. Gould.

40. Paul Sledzik, Dennis Dirkmaat, Robert W. Mann, Thomas D. Holland, Amy Zelson Mundorff, Bradley J. Adams, Christian M. Crowder, and Frank DePaola, "Disaster Victim Recovery and Identification: Forensic Anthropology in the Aftermath of September 11," in *Hard Evidence: Case Studies in Forensic Anthropology*, ed. Dawnie Wolfe Steadman (Upper Saddle River, NJ: Prentice Hall, 2009), 289–302.

41. Mark Bulliet, "More Body Parts—Surprise At WTC," *New York Post*, October 20, 2006, 21; David J. Burney and Charles J. Maikish, "Report and Recommendations on Additional Searches for Human Remains at and in the Vicinity of the World Trade Center Site," October 27, 2006, www.nyc.gov/html/om/pdf/ddc

_report.pdf; Patrick Gallahue, "Mike: Look Again; WTC Hunt is Due After Worker Finds Remains," *New York Post*, October 21, 2006, 2.

42. Bulliet, "More Body Parts."

43. Dennis McKeon, "Human Remains at Ground Zero," *Where to Turn* (blog), October 19, 2006, http://www.where-to-turn.org/phpBB2/viewtopic.php?t=31794.

44. Marcus Franklin, "Families Angry Utility Crew Finds More Human Remains at World Trade Center Site," Associated Press, October 20, 2006.

45. David W. Dunlap, "Ground Zero Forensic Team Is Posted to Seek Remains," *New York Times*, October 21, 2006, B2.

46. Gallahue, "Mike: Look Again."

47. Peter Kadushin and Paul H. B. Shin, "15 More Remains Recovered At WTC," *New York Daily News*, October 22, 2006, 7; Tina Moore, "More Ground Zero Remains Are Found. Total at 114 Pieces, But Construction Goes On," *New York Daily News*, October 23, 2006, 4.

48. Michael Daly, "His Prophecy from the Pit Comes True," *New York Daily News*, October 22, 2006, 7.

49. Charles E. Schumer and Hilary Rodham Clinton, "Schumer, Clinton Renew Call for Additional Assistance in Search for Human Remains at Ground Zero" (press release, posted on *Where to Turn*), October 20, 2006, http://www.where-to-turn.org/phpBB2/viewtopic.php?t=31838.

50. Associated Press, "World Trade Center Officials Wanted More Time to Recover Human Remains," *Fox News.com,* October 24, 2006.

51. Patrick Brennan, "Update from Mayor's Office to 9/11 Family Groups," *Where to Turn* (blog), October 26, 2006, http://www.where-to-turn.org/phpBB2/viewtopic.php?p=32052.

52. "More on Human Remains," *Where to Turn* (blog), October 22, 2006, http://www.where-to-turn.org/phpBB2/viewtopic.php?t=31901&highlight=%93more+human+remains&sid=69019ceec64518265c2ba0cf617da8ef.

53. Christine Kearney, "New York Mayor Defends Ground Zero Remains Search," Reuters, October 23, 2006.

54. Ibid.; Put It Above Ground, "New Remains Update 10/20," October 20, 2006, http://www.where-to-turn.org/phpBB2/viewtopic.php?t=31853.

55. "More on Human Remains," *Where to Turn.*

56. Put It Above Ground, "Human Remains Update 10/22" (October 22, 2006). http://www.where-to-turn.org/phpBB2/viewtopic.php?t=31910.

57. Put It Above Ground, "Human Remains Update 10/25 11:00 AM," October 25, 2006, http://www.where-to-turn.org/phpBB2/viewtopic.php?t=32072.

58. Put It Above Ground, "Human Remains Update 10/25 6:00 PM," October 25, 2006, http://www.where-to-turn.org/phpBB2/viewtopic.php?t=32109.

59. Burney and Maikish, "Report and Recommendations," 2.

60. Burney and Maikish, "Report and Recommendations."

61. New York City Fire Department, "Mayor Bloomberg Accepts Report on Expanded Search for Those Lost During Attack on World Trade Center" *NYC.gov,* October 27, 2006, http://www.nyc.gov/html/fdny/html/pr/2006/102706_7506.shtml.

62. Put It Above Ground, "Human Remains Update 10/27 10:00 PM" (October 27, 2006). http://www.where-to-turn.org/phpBB2/viewtopic.php?t=32225.

63. Put it Above Ground, "City Is Playing Russian Roulette with the Search & Our L[ives]," November 9, 2006, http://www.where-to-turn.org/phpBB2/viewtopic.php?t=32697.

64. Paul D. Colford, "Call in Military, 9-11 Kin Say. Don't Trust City to Find Remains," *New York Daily News*, November 3, 2006, 46.

65. Manhattan Community Board #1, "Minutes of the Monthly Meeting," December 19, 2006, http://www.nyc.gov/html/mancb1/downloads/pdf/FBM_Minutes/FBM_Minutes_06-12-06.pdf.

66. Patrick Brennan, letter to Mayor Bloomberg, "Search for Human Remains at the World Trade Center Site, Update 11/15," *Where to Turn* (blog), November 15, 2006, http://www.where-to-turn.org/phpBB2/viewtopic.php?t=32841.

67. Put It Above Ground, "Put It Above Ground Concerns with City Update," November 15, 2006, http://www.where-to-turn.org/phpBB2/viewtopic.php?t=32842.

68. Put It Above Ground, "They Broke Our Hearts They Will Not Break Our Spirit," November 16, 2006, http://www.where-to-turn.org/phpBB2/viewtopic.php?t=32880.

69. Diane Horning, "WTC Families for Proper Burial Update (11.28)," *Voices of September 11,* November 28, 2006, http://www.voicesofseptember11.org/dev/content.php?idtocitems=wtcfamilyupdate.

70. Put It Above Ground, "Human Remains Update 12/8 6:45 PM," December 8, 2006, http://www.where-to-turn.org/phpBB2/viewtopic.php?t=33640; Put It Above Ground, "Diane Keeps Asking and the City Keeps Ignoring Her Requests," January 4, 2007, http://www.where-to-turn.org/phpBB2/viewtopic.php?t=34106.

71. Put It Above Ground, "There Is a Method to Their Madness," December 16, 2006, http://www.where-to-turn.org/phpBB2/viewtopic.php?t=33797.

72. Ibid.

73. Jane Pollicino, letter to Governor Eliot Spitzer, *Where to Turn* (blog), January 10, 2007, http://www.where-to-turn.org/phpBB2/viewtopic.php?t=34275.

74. Put It Above Ground, "Human Remains Update 12/4 8:15 PM," December 4, 2006, http://www.where-to-turn.org/phpBB2/viewtopic.php?t=33543.

75. Edward Skyler, "Update on Search for Human Remains at the World Trade Center Site," *Where to Turn* (blog) December 29, 2006, http://www.where-to-turn.org/pdf/WTC_Search_Update_12-29.pdf.

76. Patrick Brennan, letter to 9/11 Family Groups," December 18, 2006, http://www.where-to-turn.org/phpBB2/viewtopic.php?t=33811.

77. Skyler, "Update on Search."

78. Michael Saul, "Four More Bones Found at WTC Dig Site," *New York Daily News*, December 30, 2006, 16.

79. Put It Above Ground, "Human Remains Update 1/11," January 11, 2007, http://www.where-to-turn.org/phpBB2/viewtopic.php?t=34306.

80. Dennis McKeon, "Family Tour of 11 Water Street," January 9, 2007, http://www.where-to-turn.org/phpBB2/viewtopic.php?t=34244.

81. Ibid.

82. Sally Regenhard, letter to New York City and the OCME, January 27, 2007, http://www.where-to-turn.org/phpBB2/viewtopic.php?t=34880.

83. Manhattan Community Board #1, "Minutes of the Monthly Meeting," January 16, 2007, http://www.nyc.gov/html/mancb1/downloads/pdf/FBM_Minutes/FBM_Minutes_07-01-16.pdf.

84. Ibid.

85. Skye H. McFarlane, "C.B. 1 Sides with 9/11 Families on Remains Search," *Downtown Express,* January 26–February 1, 2007, http://www.downtownexpress.com/de_194/cb1sides.html.

86. David W. Dunlap, "Search for Remains Will Go Beneath an Asphalt Lot," *New York Times*, February 1, 2007, B2.

87. Ibid.; Put It Above Ground, "Human Remains Update 1/31 No Remains 2 Steel Beams Found," January 31, 2007, http://www.where-to-turn.org/phpBB2/viewtopic.php?t=35018.

88. Put It Above Ground, "It Is Time for a Change," February 1, 2007, http://www.where-to-turn.org/phpBB2/viewtopic.php?t=35023.

89. James McCaffrey, "The Shame on Our Doorstep: On Two Crucial 9/11 Issues New York Has Failed America," *Where to Turn* (blog), February 3, 2007, http://www.where-to-turn.org/phpBB2/viewtopic.php?t=35109.

90. James McCaffrey, "To the Editor: The Forgotten Screams," *Where to Turn* (blog), April 7, 2007, http://www.where-to-turn.org/phpBB2/viewtopic.php?t=36666.

91. Put It Above Ground, "Some Things Never Change," May 5, 2007, http://www.where-to-turn.org/phpBB2/viewtopic.php?t=37204.

92. Edward Skyler, "WTC Human Remains Recovery Update," *Voices of September 11,* June 1, 2007, http://www.voicesofseptember11.org/dev/PDF/wtcupdate0601.pdf.

93. Ibid.

94. Ibid.

95. Ibid.

96. Office of Chief Medical Examiner, "PHR Reconciliation Report," July 1, 2013, www.where-to-turn.org/phpBB2/download.php?id=1857.

97. Skyler, "WTC Human Remains Recovery Update."

98. Diane Horning, "Rebuttal to June 1 Remains Recovery Update," *Voices of September 11,* June 4, 2007, https://web.archive.org/web/20100528025939/http://www.voicesofseptember11.org/dev/content.php?idtocitems=rebuttalremains.

99. Diane Horning, "Partial Response from ME. Questions Still Unanswered," June 6, 2007, http://www.where-to-turn.org/phpBB2/viewtopic.php?t=37693; Diane Horning, "Still Waiting to Hear," June 18, 2007, http://www.where-to-turn.org/phpBB2/viewtopic.php?t=37875; Diane Horning, "Can You Hear Me Now?" July 25, 2007, http://www.where-to-turn.org/phpBB2/viewtopic.php?t=38503; Put It Above Ground, "Waiting to Hear," July 9, 2007, http://www.where-to-turn.org/phpBB2/viewtopic.php?t=38276.

100. Ibid.

101. Edward Skyler, "WTC Human Remains Recovery Update," *Voices of September 11,* July 3, 2007, http://www.voicesofseptember11.org/dev/PDF/July3Memo.pdf.

102. Put It Above Ground, "Look What We Found. No Way to Run a Search Operation," July 3, 2007, http://www.where-to-turn.org/phpBB2/viewtopic.php?t=38136.

103. Edward Skyler, "Re: WTC Human Remains Recovery Update," December 11, 2007, *Families of September 11,* http://www.familiesofseptember11.org/docs/WTC%20Remains%20Recovery%20Update%2012-11-07.pdf.

104. Jarrod Bernstein and Diane Horning, "Correspondence Regarding Additional Remains," October 19, 2007, http://www.where-to-turn.org/phpBB2/viewtopic.php?t=39997.

105. Nicholas Varchaver, "The Tombstone at Ground Zero," *Fortune Magazine,* March 20, 2008, http://archive.fortune.com/2008/03/19/news/companies/ground_zero.fortune/index.htm?postversion=2008032004.

106. Charles V. Bagli, David W. Dunlap, and William K. Rashbaum, "Obscure Company Is Behind 9/11 Demolition Work," *New York Times*, August 23, 2007, A1.

107. Skyler, "Re: WTC Human Remains Recovery Update," December 11, 2007, 2.

108. Ibid., 4.

109. Stephen Goldsmith, memo to Mayor Michael R. Bloomberg." "Re: WTC Potential Human Remains Recovery Update," June 22, 2010, www.where-to-turn.org/phpBB2/download.php?id=1703.

110. Office of Chief Medical Examiner, "PHR Reconciliation Report."

8: Who Owns the Dead?

1. CBS New York, "9/11 Remains Moved," May 13, 2014, http://newyork.cbslocal.com/photo-galleries/2014/05/13/911-remains-moved/.

2. Alice Greenwald (9/11 Museum director), interview with author, January 25, 2012, National 9/11 Memorial and Museum Office.

3. 9/11 Memorial, "Summary of Outreach: Regarding Plans to Relocate the City of New York's Office of Chief Medical Examiner's (OCME's) Repository for the Unidentified and Unclaimed Remains of 9/11 Victims," https://assets.documentcloud.org/documents/97992/memorial-museum-response.pdf, 1.

4. Ibid. These groups included 9/11 Families for a Safe and Strong America, 9/11 Parents and Families of Firefighters and World Trade Center Victims, the 9/11 Widows' and Families' Association, Advocates for a 9/11 Fallen Heroes Memorial, the Cantor Fitzgerald Relief Fund, the Coalition of 9/11 Families, Families of September 11th, the FDNY Families Advisory Council, Fix the Fund, Give Your Voice, the LMDC Family Advisory Council, MyGoodDeed, Peaceful Tomorrows, September 11th Education Fund, September's Mission, the Skyscraper Safety Campaign, Take Back the Memorial, Tuesday's Children, Voices of September 11th, Windows of Hope, WTC Families for a Proper Burial, WTC United Family Group, and other relief organizations started by the corporations that lost employees on 9/11.

5. Ibid.

6. Lower Manhattan Development Corporation, "World Trade Center Site: Memorial Competition Guidelines," 2003, 19, https://www.911memorial.org/sites/all/files/LMDC%20Memorial%20Guidelines.pdf.

7. National September 11 Memorial and Museum, "Transcript of Meeting to Discuss Human Remains at the National September 11 Memorial & Museum," June 8, 2010. From the collection of the Lower Manhattan Development Corporation. On file with author.

8. National September 11 Memorial Museum, "Museum Planning Conversation Series Report 2006–2008," 2008, 2, http://www.911memorial.org/sites/all/files/Conversation%20Series%202006%20-%202008_0.pdf.

9. Ibid.

10. Ibid.

11. Ibid., 13.

12. Glenn Corbett, Sally Regenhard, and Norman Siegel, group interview with author, July 26, 2011, Norman Siegel's office.

13. National September 11 Memorial Museum, "Museum Planning Conversation Series Report," 4.

14. Edward T. Linenthal, *Preserving Memory: The Struggle to Create America's Holocaust Museum* (New York: Viking, 1995).

15. National September 11 Memorial Museum, "Museum Planning Conversation Series Report," 6.

16. Ibid., 17–18.

17. Ibid., 18–19.

18. James S. Russell, "Top-Secret $510 Million Ground Zero Museum Wallows in Grief," *Bloomberg Business,* June 1, 2007, http://wirednewyork.com/forum/showthread.php?t=4463&page=117.

19. Diane Horning and Bill Doyle, letter to Alice Greenwald, WTC Museum director, and responses, "Memorial Artifacts: Letters and Responses," *Where to Turn* (blog), June 6, 2007, http://www.where-to-turn.org/phpBB2/viewtopic.php?t=37692.

20. Ibid.

21. Ibid.

22. Ibid.

23. Jim Riches, comments, "Who Owns the Dead?" April 6, 2011, *PamelaGeller .com,* http://pamelageller.com/2011/04/911-human-remains-who-owns-the-dead.html/.

24. Ibid.

25. Greenwald gave many versions of this talk from 2007 to 2010 and eventually published an article by the same name: "Passion on All Sides: Planning a Memorial Museum at Ground Zero," *Curator: The Museum Journal* 53, no. 1 (2010): 117–125.

26. National September 11 Memorial and Museum, "Transcript of Meeting to Discuss Human Remains."

27. Native American Graves Protection and Repatriation Act [NAGPRA], Pub, L. No. 101-601, 25 U.S.C. 3001–3013, 43 CFR 10.6(a).

28. Stephen E. Nash and Chip Colwell-Chanthaphonh, "NAGPRA after Two Decades," *Museum Anthropology* 33, no. 2 (2010): 99–104.

29. Chip Colwell-Chanthaphonh, "'The Disappeared': Power Over the Dead in the Aftermath of 9/11," *Anthropology Today* 27, no. 3 (2011): 9.

30. Ibid.

31. Ibid., 10.

32. Ibid.

33. Ibid.

34. Chip Colwell-Chanthaphonh, phone interview with author, July 11, 2011.

35. Ibid.

36. Ibid.

37. National September 11 Memorial and Museum, "Transcript of Meeting to Discuss Human Remains."

38. Lower Manhattan Development Corporation, "Request for Qualifications: Exhibition Design Services," October 3, 2006. On file with author.

39. Ibid., 4.

40. Colwell-Chanthaphonh, "The Disappeared," 8.

41. National September 11 Memorial and Museum, "Transcript of Meeting to Discuss Human Remains," 9.

42. Ibid., 15.

43. Ibid., 10.

44. Ibid.

45. Ibid.

46. Ibid., 8.

47. Ibid., 11. Punctuation added for clarity.

48. Ibid., 13.
49. Ibid., 14.
50. Ibid., 19.
51. Ibid., 18.
52. Ibid., 17–18.
53. Ibid., 18.
54. Ibid.
55. Ibid., 33.
56. Ibid., 18.
57. Ibid.
58. Ibid., 19.
59. Ibid., 20.
60. Ibid., 24.
61. Ibid., 32.
62. Norman Siegel letter to Mayor Michael Bloomberg, April 5, 2011, http://www.where-to-turn.org/phpBB2/viewtopic.php?t=53884.
63. See Regenhard et al. v. City of New York 2011, No. 109548 CV, 2011 (NY Sup. Ct. Oct 25 2011) for a detailed history.
64. Corbett, Regenhard, and Siegel, group interview.
65. Regenhard et al. v. City of New York.
66. Thomas J. Meehan III, "A Sept. 11 Question: Who Owns the Dead?," July 29, 2010, http://www.where-to-turn.org/phpBB2/viewtopic.php?t=51907.
67. Joe Daniels and Charles S. Hirsch, "Letter to Family Members Regarding Location of Remains Repository at the World Trade Center Site," October 17, 2011. On file with author.
68. Ibid., 2.
69. Matter of Regenhard v. City of New York, 102 AD3d 612, (Appellate Division, January 31, 2013).
70. Caroline Alexander, "Out of Context," *New York Times*, April 6, 2011, A27.
71. Ibid.
72. David W. Dunlap, "A Memorial Inscription's Grim Origins," *New York Times*, April 2, 2014, A20; David W. Dunlap, "Scholarly Perspectives on Inscription at the 9/11 Memorial Museum," *New York Times*, April 2, 2014, A20.
73. Peter Kruschwitz, "Misappropriation and Misapprehension: Vergil on 9/11," *The Petrified Muse,* April 4, 2014, https://thepetrifiedmuse.wordpress.com/2014/04/04/misappropriation-and-misapprehension-vergil-on-911/.
74. Dunlap, "A Memorial Inscription's Grim Origins."
75. Virginia Bauer, Paula Grant Berry, Mary Fetchet, Frank Fetchet, Christine A. Ferer, Monica Iken, Anthoula Katsimatides, Margie Miller, Tom Rogér, and Charles Wolf, letter to 9/11 family members, April 5, 2011, *Where to Turn* (blog) http://www.where-to-turn.org/phpBB2/viewtopic.php?t=53881.

76. Monica Iken, Thomas S. Johnson, and Charles Wolf, "How to Honor the Lost at Ground Zero," *Wall Street Journal*, April 20, 2011, A15.

77. Ibid.

78. Sarah Wagner, "The Making and Unmaking of an Unknown Soldier," *Social Studies of Science* 43, no. 5 (2013): 631–656.

79. Colwell-Chanthaphonh, interview.

80. Iken, Johnson, and Wolf, "How to Honor."

81. Maureen Santora and Al Santora, "9/11 Parents Outraged at Set the Record Straight Statement," *Where to Turn* (blog), April 8, 2011, http://www.where-to-turn.org/phpBB2/viewtopic.php?t=53898.

82. Thomas J. Meehan III, letter to Mary Fetchet, posted as "Concerns about the Remains /a Father's Response," *Where to Turn* (blog), April 6, 2011, http://www.where-to-turn.org/phpBB2/viewtopic.php?t=53886.

83. Ibid.

84. Charles V. Bagli, "Dispute Over Costs Delays Opening of 9/11 Museum," *New York Times*, September 8, 2012, A1.

85. Monika Iken, interview with Caroline Weaver, "September 11 Museum in New York Still Unfinished," *Voice of America*, September 5, 2012, http://www.voanews.com/content/national-911-museum-at-standstill-as-anniversary-nears/1502546.html.

86. Port Authority of New York and New Jersey, and National September 11 Memorial and Museum, "Port-Memorial Memorandum of Understanding," *NYC.com*, September 10, 2012, www.nyc.gov/html/om/pdf/2012/911_museum_mou.pdf.

87. David W. Dunlap, "Floodwater Pours into 9/11 Museum, Hampering Further Work on the Site," *New York Times*, November 2, 2012, A19.

88. David W. Dunlap, "Risk of Flooding Will Not Alter Plan to Preserve 9/11 Remains," *New York Times*, February 21, 2013, A19.

89. Ibid.

90. President Barack Obama, "Remarks by the President at 9/11 Museum Dedication," whitehouse.gov, May 15, 2014, http://www.whitehouse.gov/the-press-office/2014/05/15/remarks-president-911-museum-dedication.

Epilogue

1. Caroline Bankoff, "9/11 Museum Not the Best Place for a Cocktail Party," *New York Magazine*, May 21, 2014, http://nymag.com/daily/intelligencer/2014/05/911-museum-not-the-best-place-for-a-party.html.

2. CBS New York, "9/11 Families Blast 'Greedy,' 'Disrespectful' Decision to Store Unidentified Remains Underground in Museum," May 8, 2014, http://newyork.cbslocal.com/2014/05/08/911-families-blast-greedy-disrespectful-decision-to-store-unidentified-remains-underground-in-museum/.

3. Ibid.

4. See, for example, Charles B. Strozier and Scott Gabriel Knowles, "How to Honor the Dead We Cannot Name," *Slate*, May 12, 2014, http://www.slate.com/articles/health_and_science/science/2014/05/september_11_memorial_museum_controversy_unidentified_remains_and_lessons.html.

5. Barbara A. Sampson, letter to families and friends, May 2014, http://tributewtc.org/wp-content/uploads/2014/05/OCMELetter.pdf.

6. Strozier and Knowles, "How to Honor the Dead."

7. Kevin Conlon, "Mom of 9/11 victim: Identified remains 'finally put everything to rest'," *CNN.com,* March 20, 2015, http://www.cnn.com/2015/03/20/us/9-11-remains-identified/.

8. Ibid.

Acknowledgments

In many ways, this is a book I never expected to write. In the course of my research on the identification of the missing after conflict and disaster, I knew I would have to engage with the World Trade Center terrorist attacks in some way, but I did not want to become mired in the political and cultural controversies surrounding September 11th. My initial plan was to write two scholarly articles—one on the forensic recovery and identification effort and another on the storage of human remains—and then move on. The further I got into the details, however, the clearer it became that I could do justice to this story only in book form. I then set out to produce a narrative that is compassionate to the victims and their families, and as sympathetic as possible about the motives of the parties involved in the recovery, identification, and memorialization efforts. I hope that readers, and especially those affected by the terrorist attacks and their aftermath, agree that I achieved this goal.

Along the way, I have received the support, guidance, and assistance of numerous individuals, organizations, and institutions. Carnegie Mellon University, and especially my colleagues in the History Department, have provided the support and freedom necessary for me to give the material the attention it deserved. I especially want to thank Caroline Acker, John Lehoczky, Alex John London, Richard Scheines, and Joe Trotter for their professional goodwill and guidance. Leslie Levine, my grants administrator, has provided sage advice, financial acumen, and good humor in abundance.

This project was supported by a generous grant from the U.S. National Institutes of Health to study the ethical and political dimensions of postconflict and postdisaster identification of the missing (Grant #R01HG005702). Special thanks go to my amazing program officer, Jean McEwen, for all of her assistance and enthusiasm over the years. My collaborator on this grant, Sarah Wagner, has been a continued intellectual inspiration for me in this field. Her work has been central to my understanding of the social, cultural, and political dimensions of the treatment of human remains after conflict and disaster. She provided critical feedback on portions of this project and forced me to think more deeply not just about the practice of forensic work but also about its meaning and cultural significance. I would also like to thank Adam Rosenblatt, Lindsay Smith, and Lola Vollen for their participation in the project and their intellectual engagement over the past five years.

I am also grateful to numerous participants in the story who shared their experiences and provided materials relevant to my research. First and foremost, Amy Mundorff's generous gift of her time and engagement knew almost no bounds. I am deeply appreciative of her painstaking efforts to help me understand the forensic investigation, and her willingness to read early drafts and provide critical comments. Diane Horning spent many hours with me explaining her views on the handling of World Trade Center remains. She also provided invaluable feedback on my manuscript. Sally Regenhard showed me around the World Trade Center site and the surrounding neighborhood when I was just beginning my research. She spent many hours on the phone with me discussing her experiences as the mother of a firefighter who died in the attacks as well as her activism since September 11, 2001. Alice Greenwald and Michael Frazier spoke to me at length about the planning and construction of the memorial and museum and gave me a tour of the museum space before completion. Mark Schaming shared his experiences observing the recovery effort at Fresh Kills and provided access to the archives of the New York State Museum relating to the recovery effort. Chip Colwell-Chanthaphonh discussed the ethical and political dimensions of human remains storage in anthropological contexts and shared his experiences of the debate over the handling of World Trade Center remains.

Many thanks to the numerous people who spoke to me during the course of my research, including Brad Adams, Lynn Castrianno, Glenn Corbett, Charles Flood, Rosemary Foti, Anthony Gardner, Richard Gould, Gordon Haberman, Mike Hennessey, Alice Hoagland, Monica Iken, Arnie Korotkin, Thomas Meehan, Anne Papageorge, Mark Perlin, Shiya Ribowsky,

Robert Shaler, Norman Siegel, Amanda Sozer, Giovana Vidoli, and others with whom I met informally or off the record. I would be remiss not to mention the numerous authors, journalists, lawyers, oral historians, and scholars who have documented various aspects of this story and made their findings and materials available for public use. The Lower Manhattan Development Council maintains a sizeable Internet archive of documents related to the redevelopment of the World Trade Center site and provided me with access to transcripts of meetings that were not available online.

At Harvard University Press, I have benefited a great deal from my editors' feedback. Thanks especially to Elizabeth Knoll for believing in this project in its early stages and for helping me find my authorial voice. Andrew Kinney deserves much praise for picking up the project halfway through development and shepherding it through the review and production process. In addition to the press's anonymous reviewers, numerous others have read this work in various stages and helped me make it better. Thanks go to Lea David, whose critical reading of my introduction and conclusion brought much greater clarity and sharpness to both. Ed Linenthal provided the kind of feedback that an author dreams of receiving from a reviewer: criticism leavened with praise. He also provided numerous references that allowed me to expand the theoretical and scholarly reach of my work. Although she was not directly involved in this project, I owe a tremendous debt of gratitude to Sheila Jasanoff for her ongoing support of me as both a scholar and a person.

Finally, I would like to thank my family—especially my wife, Tamara; my children, Ezra, Talia, and Maayan; the Heilmans; and the Dubowitzes—for keeping me grounded and reminding daily of what it means to love and be loved. I thought about them often during the writing of this book, both as a reminder of the emotional pain that the families of the victims must have felt and as a reason to always endeavor to make the world a better place, however small my contribution may be. This book was written in their honor. It is dedicated to the memory of the victims of the World Trade Center attacks and the untold number of victims of the war, terrorism, and violence that have emerged in their wake.

Index